TEXTBOOK OF PETROLOGY

Volume Three

Petrology of the Metamorphic Rocks

TITLES OF RELATED INTEREST

Rutley's elements of mineralogy, 26th edn
H. H. Read

Petrology of the igneous rocks, 13th edn
F. H. Hatch, A. K. Wells & M. K. Wells

Metamorphism and metamorphic belts
A. Miyashiro

Petrology of the sedimentary rocks, 6th edn
J. T. Greensmith

Metamorphic processes
R. H. Vernon

The interpretation of igneous rocks
K. G. Cox, J. D. Bell & R. J. Pankhurst

The inaccessible Earth
G. C. Brown & A. E. Mussett

Tectonic processes
D. Weyman

Sedimentology: process and product
M. R. Leeder

Introduction to small-scale geological structures
G. Wilson

Metamorphic geology
C. Gillen

The poetry of geology
R. M. Hazen (ed.)

Komatiites
N. T. Arndt & E. G. Nisbet (eds)

Statistical methods in geology
R. F. Cheeney

A practical introduction to optical mineralogy
C. D. Gribble & A. J. Hall

Rock mechanics for underground mining
B. Brady & E. T. Brown

Hemispherical projection methods in rock mechanics
S. D. Priest

Petroleum geology
F. K. North

Petrology of the Metamorphic Rocks

ROGER MASON
University College, University of London

London
GEORGE ALLEN & UNWIN
Boston   Sydney

George Allen & Unwin (Publishers) Ltd,
40 Museum Street, London WC1A 1LU, UK

George Allen & Unwin (Publishers) Ltd,
Park Lane, Hemel Hempstead, Herts HP2 4TE, UK

Allen & Unwin Inc.,
9 Winchester Terrace, Winchester, Mass 01890, USA

George Allen & Unwin Australia Pty Ltd,
8 Napier Street, North Sydney, NSW 2060, Australia

First published in 1978
Fourth impression 1984

British Library Cataloguing in Publication Data

Mason, Roger
   Petrology of the metamorphic rocks:—
(Textbook of petrology; vol. 3).
1. Rocks, Metamorphic
I. Title   II. Series
552'.4   QE475.A2   78-40124
ISBN 0-04-552013-5
ISBN 0-04-552014-3 Pbk.

Cover photograph by Mike Gray
Typeset in 10 on 12 point Times by George Over Limited, London and Rugby
and printed in Great Britain
by Mackays of Chatham Ltd

To S. O. Agrell and G. A. Chinner

Foreword

With the appearance of *Petrology of the metamorphic rocks* a long-standing wish of mine has been satisfied, namely the publication of a book on metamorphic rocks designed primarily for non-specialist students interested in the Earth sciences, and which would in this way form a companion to the existing books concerned with igneous and sedimentary rocks.

Individual authors necessarily approach their subjects differently not only because of their varied interests but also because of major differences in subject matter. Igneous rocks of a volcanic nature can actually be observed in the making, albeit from a respectful distance, so it has been natural that in studying igneous rocks attention has been directed very much to laboratory investigations into the crystallisation of minerals from silicate melts. Similarly, with sedimentary rocks it is generally possible to gain critical insight into their origins by observing present conditions of sedimentation and by exploration of the oceans. Metamorphic processes, however, can never be observed directly. They are largely, therefore, matters of inference from evidence provided by the rocks themselves, backed by the accumulated evidence of field studies which may be regarded as case histories relating to given sets of geological conditions and events. Since both the nature of the rocks prior to metamorphism, and the conditions to which they are subjected, vary very widely, it is impossible to do justice to all in one book, and certainly in one of an introductory nature. This is perhaps the main justification for emphasis in the present case on selected topics which have proved of special interest to the Author in his teaching and research experience.

Certainly from my own experience there are few pursuits more exciting and rewarding than studying crystalline rocks, ideally in the field in the first instance, and then petrographically, in order to determine their nature and to interpret their histories. Fortunately this is an experience in which anyone can share who is prepared to gain acquaintance with the minerals concerned and to learn the basic rules of petrographic interpretation.

Although a mainly petrographic approach has thus been adopted for this book — as for its igneous and sedimentary companions — other considerations vital to an understanding of metamorphic rocks are not neglected. Improvements in experimental techniques have allowed minerals stable in metamorphic rocks to be investigated with an intensity and degree of precision previously reserved for those of volcanic rocks. Great strides have also been made in mineral analysis with instruments like the electron microprobe, to which the reader is introduced. Finally, the study of trace-element distribution, particularly of radiogenic isotopes, has provided students of crystalline rocks with tools which no modern Earth scientist can afford to neglect: they can reveal evidence of ultimate origins of rock materials from the crust or mantle, and of the timing and rates of geological processes.

We are finding increasingly that rocks can retain remarkably complete memories of the conditions and events which have affected them. The starting point in releasing these memories is careful examination of the rocks in the field, with a lens or a microscope. It is here that the reader will hopefully find this book of value.

M. K. Wells

Preface

The writing of this book has been urged upon me by teaching colleagues in universities and polytechnics, by the publishers and not least by my students. There is no textbook more recent than A. Harker's classic work (which is nearly half a century old) which treats metamorphic rocks on their own from a petrographic rather than a petrogenetic viewpoint. There is also a need for a book above the introductory level, but comprehensible by geology students who are not petrology specialists. Such students are often alarmed by the physical chemistry involved in the theory of metamorphic petrogenesis, and come to regard metamorphic rocks as peculiarly complex and difficult to understand. I hope I can help to enlighten them.

This book has evolved from a course of lectures given to second-year students in a three-year honours B.Sc. course in geology. The method adopted is to illustrate general principles from the evidence presented by selected suites of metamorphic rocks, particularly when studied under the petrological microscope. This inevitably means that some subjects are treated in greater depth than others, and some have probably received too little attention. It also means that general principles are expounded in a rather haphazard order throughout the text. Therefore to help the reader a glossary of definitions of the terms used in the book is provided. Terms in the glossary are indicated in **bold type** when they first appear and also subsequently where the context suggests that the reader might wish to refresh his memory of the definition.

My theme throughout is the recognition of assemblages of minerals coexisting in equilibrium during metamorphism. The substance of the book is in Part II where I try to illustrate the criteria for recognising such assemblages in a series of examples. I also discuss briefly the way in which estimates of temperature, pressure and activity of volatile phases during metamorphism may be derived from the study of mineral assemblages. I have given less prominence than some textbooks to the definition of a scheme of metamorphic facies as a way of interpreting mineral assemblages, believing that in doing so I am following the tendency among metamorphic petrologists today. I have been more concerned to show how ideas on metamorphic petrogenesis are founded in the study of the rocks themselves than to give a comprehensive account of those ideas. I hope I have thereby provided the reader with a sound basis for further exploration of this fascinating and rapidly advancing branch of the Earth sciences.

A large number of people have helped me during the preparation and writing of this book. I must especially thank the following, who supplied me with rocks to describe, and improved my understanding of branches of the subject of which I was ignorant: Dr S. O. Agrell, Dr R. Hall, Dr B. Harte, Professor R. A. Howie, Dr R. M. F. Preston, Dr S. M. F. Sheppard, Dr J. D. Smewing, Dr C. Taylor, Dr J. H. Milledge and Dr M. K. Wells. If I have misinterpreted what they told me, the fault is mine, not theirs. Many of my students have offered valuable advice from the consumers' viewpoint, notably members of the class who graduated in geology from University College in 1976. My views on metamorphism have been strongly influenced by my having read Professor A. Miyashiro's book on metamorphism before its publication, and corresponded with him about it. Dr S. W. Richardson, Professor Janet Watson and Dr R. Vernon have read the manuscript, correcting numerous errors and suggesting many improvements. I would like to thank Dr P. W. Edmondson for help in preparing the index, and my wife Joy for her help and support throughout the writing of this book. Finally, I must thank Roger Jones of George Allen & Unwin, without whose tactful and patient persuasion I would never have completed this work.

Roger Mason,
Department of Geology
University College London
September 1977

Contents

Tables

Part I
Metamorphic rocks in the field

1

Metamorphism and metamorphic rocks

DEFINITION

Metamorphic rocks are those whose character has been changed since their original formation by processes operating within the Earth (or other planetary bodies). The change may include a change in the minerals making up the rock, or a change in the relationship between those minerals, the **texture** of the rock, or both. 'Metamorphism' is a general name given to the processes of change. All metamorphic rocks were once igneous or sedimentary, although the processes of metamorphism may have changed them so much that their original nature is unrecognisable.

Metamorphism almost always involves the partial or complete recrystallisation of the original rock minerals. The original crystals of the rock have been broken down and new crystals have grown. The recrystallisation has taken place at temperatures below those at which the rock melts. Thus metamorphism involves processes of *solid-state* recrystallisation. Geological observation and laboratory experiments in solid-state recrystallisation (see Ch. 13) indicate that recrystallisation occurs much more rapidly as temperature increases. The temperature and pressure during metamorphism, and the composition of the pore-fluid (if any) in the rock, influence the minerals present in the metamorphosed rock.

It is not sufficient to define metamorphic rocks as those which have recrystallised since their original sedimentary deposition or igneous crystallisation. Many sedimentary rocks have undergone recrystallisation after their primary deposition, a process which geologists call **diagenesis**. This process falls outside the generally accepted scope of metamorphism, because it may affect sediment which is buried only a few metres below the surface, and may occur very soon after deposition. The distinction between diagenesis (which is therefore regarded as a sedimentary process) and metamorphism is conventionally drawn by saying that diagenesis takes place under conditions of temperature, pressure and ground water composition comparable with those found at the surface of the Earth's crust, whereas metamorphism involves higher temperatures or pressures, or different pore-fluid compositions, or combinations of these. This is a genetic distinction, difficult to apply in practice to rocks on the borderline between the two processes.

Igneous rocks may undergo solid-state recrystallisation after they have crystallised from molten magma but before they have cooled to the temperature at the surface of the Earth's crust. For example, glassy lavas may partially devitrify, or granitic rocks recrystallise locally with hydrous minerals such as muscovite, kaolinite and epidote replacing feldspars. Rocks formed by such **deuteric alteration** are usually not regarded as metamorphic rocks. In some cases, for example where deuterically altered igneous rocks have been intruded while their **country rocks** were being metamorphosed, metamorphism and deuteric alteration are indistinguishable. However, for the great majority of metamorphic rocks, a brief examination of a hand-specimen is sufficient to show compositional and textural features which differ from those of sedimentary or igneous rocks.

The definition of metamorphic rocks may be summarised as follows. Metamorphic rocks have undergone partial or complete recrystallisation under conditions of temperature and pressure significantly higher than those found at the surface of the Earth's crust, but below the temperature at which they melt. Note that in this definition the phrase 'at the surface of the Earth's crust' includes the range of conditions found on the surface of the land and also on the beds of the seas and oceans.

THE THREE FIELD CATEGORIES OF METAMORPHIC ROCKS

Metamorphic rocks are divided into three categories on the basis of their field occurrence: **contact** metamorphic rocks, **dynamic** metamorphic rocks and **regional** metamorphic rocks. The first two categories include rocks which outcrop over small areas in particular geological settings. Rocks in the third category outcrop over large areas and in a variety of settings.

Contact metamorphic rocks occur at or near the contacts of igneous intrusions. In some cases the degree of metamorphic recrystallisation can be seen to increase as the surface of contact between the country rocks and the igneous intrusion is approached (Ch. 2 and Ch. 5). This relationship suggests that the most important agent causing metamorphic recrystallisation in these rocks is the heat supplied to the country rock by the cooling intrusion. Therefore the process of metamorphism in contact metamorphic rocks is often called **thermal metamorphism**.

Dynamic metamorphic rocks are found in narrow zones such as major faults and thrusts, where particularly strong deformation has occurred. They also occur near the impact sites of large meteorites.

Regional metamorphic rocks occur over large tracts of the Earth's surface. They are not necessarily associated with either igneous intrusions or fault or thrust belts, although these are often present. The regional metamorphic rocks extend beyond the immediate vicinity of the intrusions, faults or thrusts, and the process of metamorphism is not obviously related to them (Ch. 9). Regional metamorphic rocks characteristically have distinctive textures, such as **schistosity** and **lineation** (Ch. 2). It is often possible to demonstrate that regional metamorphic rocks underwent metamorphism at about the time they were intensely deformed and that this deformation played a major part in their textural evolution.

AIMS OF STUDY OF METAMORPHIC ROCKS

One important aim of the study of metamorphic rocks is to attempt to discover from their present mineralogical compositions and textures the history of heating, deformation and other processes involved in the metamorphism of rocks of different areas. In attempting this, the geologist studying metamorphic rocks is at a severe disadvantage compared with his colleagues who study sedimentary and igneous rocks. They have the chance to study directly the processes of formation, of some rocks at least, in operation on the Earth at the present day, and to apply the **principle of uniformitarianism** to the formation of comparable rocks in earlier geological periods. The geologist studying metamorphic rocks has to rely almost entirely upon indirect methods to understand metamorphic processes. These include reconstruction based upon interpretation of present field relationships, minerals and textures, and laboratory investigations relating to the stability of metamorphic minerals.

For this reason the **petrogenesis**, or mode of origin of metamorphic rocks, is a matter of more uncertainty and differences of opinion than for sedimentary or igneous rocks. There is not only argument about the petrogenesis of individual suites of metamorphic rocks, but also disagreement about which features of the rocks are most relevant in the discussion of their metamorphism. Some geologists emphasise the mineral assemblages of the rocks, which may be used in conjunction with laboratory studies of mineral stability to make generalisations about the temperatures and pressures of recrystallisation. Others emphasise the study of the textures and structures of metamorphic rocks, which may be compared with structural geological data to determine the deformation and recrystallisation history of a particular area of metamorphic rocks. Although there is no fundamental contradiction between these two approaches, their proponents quite often disagree about the metamorphic processes operating in a particular area.

In recent years, there has been an upsurge of interest in metamorphic rocks. This has been stimulated by the achievement of some reliability in determining the temperatures of metamorphism, linked with at least roughly accurate estimates of pressures. Thus the study of metamorphic rocks can now reveal the distribution of temperature with depth in the geological past in parts of the Earth's crust and even upper mantle. This approach was first demonstrated by Miyashiro (1961) who discussed the evolution of island arcs of the Japanese islands, and has become incorporated into the body of evidence used to support the plate tectonic theory. More recently studies of the metamorphism of ancient oceanic rocks (Gass & Smewing 1973) and of older orogenic belts such as the Caledonides of Scotland (Dewey & Pankhurst 1970, Winchester 1974) have also been used to argue in support of modern tectonic theory. The most ancient rocks so far discovered are metamorphic and igneous, and study of metamorphic rocks is crucial to discussion of the early evolution of the Earth's crust (Read & Watson 1975a).

Metamorphic rocks are not only of academic interest. Many of the Earth's economically exploited resources of metals are found among metamorphic rocks and many are metamorphic rocks themselves (e.g. Davies 1969). Exploration for such metal deposits, and extraction once the deposits are discovered, demand a knowledge of metamorphic processes.

Thus while the study of metamorphic rocks used to be the speciality of a small group of petrologists, most students of Earth science are now aware at least of some of the conclusions derived from the study of metamorphism.

THE AIMS OF THIS BOOK

It is the aim of this book to give some indication of the observations, experiments and theoretical discussions upon which conclusions of wider value to Earth science may be founded. In the first two Parts, the emphasis will be upon observations of the type students can make for themselves. The observations of Part I are those which can be made in the field or by examining hand-specimens. Those in Part II include as well as field observations, descriptions of thin sections of rocks, studied by means of the petrological microscope. Part III discusses laboratory techniques and their results which are important in the study of metamorphic rocks.

Part II describes only a small selection of metamorphic rocks. They are mainly of two compositional types: **pelitic** sediments and basic igneous rocks (Ch. 3). The important case of progressive contact metamorphism of siliceous carbonate rocks is also briefly discussed. The two compositional types most frequently discussed are common in most types of progressive metamorphic sequence. Pelitic rocks show an unusually wide variety of changes in their minerals with varying metamorphic grade. The particular areas of metamorphic rocks selected for detailed description have been chosen because of their regional tectonic settings, as described in Chapter 2. Also, in as many cases as possible, the author has chosen rocks of which he has had some direct experience in the field or at least in demonstrating in the laboratory to student classes. This means that metamorphic rocks of classical areas such as the Barrow Zones of the Highlands of Scotland and the Abukuma Plateau of Japan, are only mentioned in passing. It is hoped that the examples described are sufficiently representative to help the student understand the metamorphic rocks of other areas and that the absence of descriptions of the rocks of classical metamorphic areas will encourage him to read the published accounts of them for himself.

2

Field relations of metamorphic rocks

In this chapter and the next, aspects of metamorphic rocks in the field will be described. The features discussed are those visible in outcrops, or in hand-specimen using a hand-lens if necessary, or those shown on geological maps. In accordance with the aims of the study of metamorphic rocks outlined in Chapter 1, two aspects of the field study of metamorphic rocks will be selectively discussed. In Chapter 2 features relevant to the processes of metamorphism of the rocks will be described, the three-fold division of metamorphic rocks into categories based upon their field relationships being elaborated and illustrated by examples. In Chapter 3 the description and nomenclature of metamorphic rocks will be discussed.

CONTACT METAMORPHIC ROCKS

Along each contact of a dolerite dyke with sedimentary rocks there is usually a zone of 1–10 mm wide over which the **country rock** has been metamorphosed. For a dyke 1 m thick there might typically be a zone 2 mm wide of fine-grained rock, whose broken surfaces resemble the broken surfaces in pottery (Fig. 2.1). Since the dolerite magma was much hotter than the country rock at the time of intrusion, perhaps 1000 °C compared with 30 °C, the alteration has obviously been caused by the strong heating of the contact zone at the time of intrusion. The narrow zone of porcellanous rock is often called the 'baked margin' of the country rock, corresponding to the 'chilled margin' of the dyke.

Thicker intrusions may have wider baked margins. The Whin Sill of northeastern England is composed of dolerite and is up to 73 m thick. There is a baked margin up to 40 m wide above and below the sill (Robinson 1972). The country rock of the baked margin is harder than the unmetamorphosed sediments, massive and fine-grained. It tends to have a conchoidal fracture

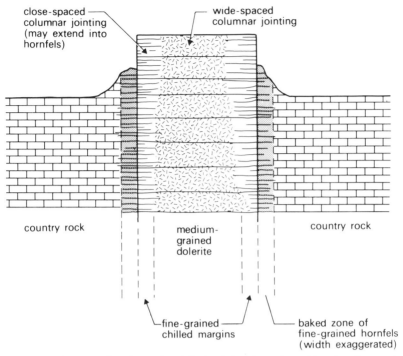

close-spaced
columnar jointing
(may extend into
hornfels)

wide-spaced
columnar jointing

country rock

medium-
grained
dolerite

country rock

fine-grained
chilled margins

baked zone of
fine-grained hornfels
(width exaggerated)

Figure 2.1 Cross section through a dolerite dyke about 1 m wide, showing contact metamorphic alteration of country rock.

and to shatter into numerous sharp-edged fragments, like flint. This type of massive rock found at the contacts of igneous intrusions is called **hornfels**.

In larger intrusions still, the baked margins also become wider. The Sulitjelma Gabbro, northern Norway, is an irregular sheet-like body reaching a maximum thickness of more than 2000 m (Mason 1971). It has a baked margin of massive hornfels 30–100 m wide, and is intruded into regional metamorphic rocks (Ch. 8). The hornfelses are readily distinguished in the field from the regional metamorphic rocks. They are relatively fine-grained and massive, whereas the regional metamorphic rocks are coarser-grained and often split easily into sheets because of the preferred parallel orientation of phyllosilicate flakes. The hornfelses are more resistant to erosion than the gabbro and the country rocks and therefore in many places a ridge marks the position of the contact. This is quite a common feature of large intrusions surrounded by hornfelses. Although the hornfelses are finer-grained than the country rocks, they are coarser-grained than those at the contacts of the Whin Sill. They can be seen to have a rather uniform grain size, giving a granular appearance to freshly broken surfaces under a hand-lens. This is called **granoblastic texture**, and is the usual texture of hornfelses (Ch. 5) although it may also be found in certain regional metamorphic rocks.

The tendency described above for the grain size of hornfelses to increase with the size of the intrusion is a general one, although there is a lot of variation in the grain size of hornfelses both in different places around any particular intrusion and between different intrusions of the same size. Large zones of contact metamorphic rocks surrounding intrusions are called **contact aureoles**. The name is applied to any zone of contact metamorphic rocks that is wide enough to be represented on a geological map.

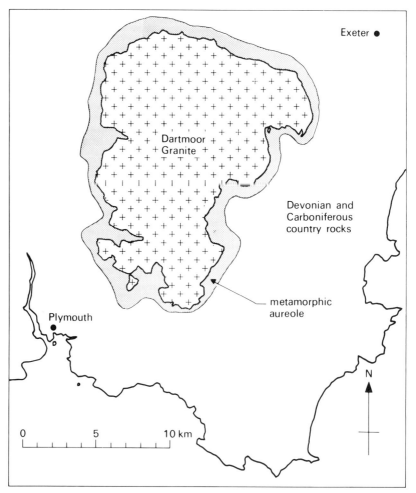

Figure 2.2   The contact aureole surrounding the Dartmoor Granite, Devon, England. Based upon Geological Survey ¼-inch sheet 22 and 1-inch sheets 324, 338, 339 and 349.

Contact metamorphic rocks are found surrounding granite and granodiorite intrusions as well as dolerite and gabbro intrusions. In small intrusions such as dykes and sills the hornfels zones are the same as those of basic intrusions. The slightly lower temperature of intrusion of acid magma does not produce a detectable difference in the width of the baked zones. Large granite and granodiorite intrusions, however, commonly have wider contact aureoles than basic intrusions of comparable size. For example, the granites of the west of England are surrounded by contact aureoles up to 2 km wide (Fig. 2.2). An even wider contact aureole surrounds the Skiddaw Granite, Cumbria (Fig. 5.1). In wide contact aureoles it is possible to recognise a sequence in the metamorphism of country rocks, metamorphic changes becoming more pronounced as the contact is approached. The sequence from unmetamorphosed country rock to the most metamorphosed hornfelses right against the intrusion is called a **progressive metamorphic sequence**. Particularly impressive sequences of textural and mineralogical changes are seen in progressive contact metamorphic sequences in which the country rock is a shale or slate. Rocks of this composition are called **pelitic rocks** by metamorphic petrologists (Ch. 3).

A good example of such a sequence in pelitic rocks is seen in the contact aureole of the Markfield Diorite in Cliffe Hill Quarry, Leicestershire (Evans *et al.* 1968). The country rock is late Precambrian **slate** of the Woodhouse Eaves Series. Diorite and hornfels are quarried for road metal, serving equally well for this purpose. Consequently, the contact is exposed in the working faces and floor of the quarry, and contact relationships can be examined in unweathered rocks over a distance of 40–50 m. The relationships of the different rock types are shown schematically in Figure 2.3.

The first recognisable change due to contact metamorphism is a change in the Woodhouse Eaves beds from slaty to massive. This is not seen in the quarry itself. Within the quarry, the rocks become more massive as the contact is approached. Between about 5 m and 25 m from the contact, dark spots appear in some bands in the rocks. They are concentrated into bands which appear to be original bedding surfaces. In the last 5 m up to the contact the rock is very hard and massive and has become a true hornfels. It shows a banding from light to dark grey, but the spots are not present. The banding is obviously sedimentary bedding, because original sedimentary structures such as channels and ripple marks may be found. Rain pits and impressions of the Precambrian fossil *Charniodiscus* have been found preserved on bedding planes in this massive hornfels zone (Evans *et al.* 1968). The sequence may be summarised:

slate   → indurated slate → spotted hornfels → massive hornfels → diorite
(country (progressive metamorphic sequence of contact aureole) (intrusion)
rock)

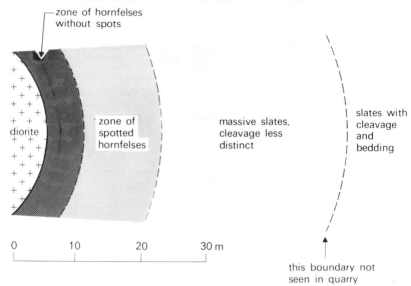

Figure 2.3 Schematic plan of the contact metamorphic zones seen in Cliffe Hill Quarry, Markfield, Leicestershire.

The unusual degree of preservation of primary sedimentary features in hornfelses close to the contact has been recorded in other contact aureoles (Mercy 1965, p. 234). The development of spots is a characteristic feature of contact metamorphic rocks of pelitic composition. The mineralogical nature of the spots varies in different contact aureoles (cf. Ch. 5) and those at Cliffe Hill appear to be graphite-rich, formed by the aggregation of organic material.

The development of the progressive metamorphic sequence in the contact aureole of the Markfield Diorite may be ascribed to increasing temperature during metamorphism as the contact is approached. It may be said that the hornfelses against the contact have undergone more intense metamorphism than those further away. Alternatively one can say that the hornfelses near the contact have a higher **metamorphic grade** than those further away. In the field, metamorphic grade may be used in a loose sense as a relative term for describing the intensity of metamorphism in rocks of progressive metamorphic sequences, 'high grade' describing the more intensely metamorphosed members of the sequence, 'low grade' the less intensely metamorphosed members. Grade may be used this way to describe the metamorphism of rocks in contact, dynamic and regional metamorphic rocks.

Although in contact metamorphic rocks a higher grade of metamorphism implies a higher temperature of metamorphism, this is not so in all progressive metamorphic sequences of dynamic and regional metamorphic rocks. The loose definition of metamorphic grade given above is adequate for discussing metamorphic rocks in the field, but when the minerals of a metamorphic rock can all be identified using the petrological microscope, a more precise definition becomes possible, and will be proposed in Chapter 8. In contact metamorphic aureoles it is possible to define stages in the progressive sequence in the field by noting the incoming of new minerals, and this will be described in Chapter 5. However, contact aureoles are more usually subdivided into zones by textural changes in the rocks, as with the Markfield Diorite aureole described above (Fig. 2.3) and with the Skiddaw aureole, Cumbria (Fig. 5.1). Comparisons of grade between one contact aureole and another are difficult to make, demanding a more complete knowledge of metamorphic minerals than can usually be obtained by field study alone.

In the inner parts of some contact aureoles, the country rocks have not only been recrystallised during metamorphism, they have also undergone a change in chemical composition. In all contact aureoles, the volatile components of sedimentary country rocks such as H_2O and CO_2 tend to decrease in metamorphosed sedimentary rocks with increasing metamorphic grade. But in some aureoles the proportions of less volatile components such as potassium, calcium or boron are also changed. Metamorphism involving a change in chemical composition of this sort is called **metasomatism**, and in cases where it occurs near the contacts of igneous intrusions it is called **contact metasomatism**.

Contact metasomatism is particularly associated with granite intrusions in which the granite magma was originally rich in H_2O and other volatiles. This volatile portion of the magma was driven off towards the end of magmatic crystallisation of the granite, after solidification was complete. It became enriched in those chemical elements that were not readily incorporated into the minerals of the granite. As this H_2O-rich fluid with a variety of metallic and non-metallic elements in solution escaped through fractures in the country rock, and in the later stages of crystallisation through the outer parts of the granite itself, the rocks alongside the fractures were not only recrystallised, their chemical composition was changed.

Such contact metasomatism is found around the granites of south-western England. The type of rock produced depends on the chemical element or elements which were introduced to and removed from the rocks. For example, many Cornish granites have undergone metasomatism near their contacts in which boron was introduced into the original granite (Hatch *et al*. 1972, p. 225). First of all the biotite in the granite was replaced by the boron-bearing mineral tourmaline, then feldspar was replaced by tourmaline and quartz. Rocks consisting entirely of tourmaline were the final products of this boron metasomatism. In general, increasing metasomatism tends to reduce the number of minerals in the contact metasomatic rocks. Rocks with one, two or three minerals formed by contact metasomatism are called **skarns** by some authors. More often, the term is restricted to calcium-rich rocks formed by metasomatism of limestones and marbles. Skarns are often associated with metallic ore deposits of hydrothermal origin.

The tendency for granites to have wider contact aureoles than gabbros has already been mentioned. There is also a considerable variation in the widths of contact aureoles round granites themselves. The northern granite of the Isle of Arran, Scotland, is approximately the same size as the Land's End Granite of Cornwall, which has an aureole as wide as that of the larger Dartmoor Granite (Fig. 2.2). The maximum width of the aureole of the Arran Granite, however, is only about 180 m (Macgregor 1972). The Arran Granite does not have contact metasomatic rocks and hydrothermal mineral veins near its contacts. It appears that the width of the contact aureole around a granite depends not only on the temperature of the magma but also on the amount of fluid given off by the granite in its late stages of crystallisation. This has important implications for the understanding of contact metamorphism. The rocks of a contact metamorphic aureole were probably not heated mainly by the conduction of heat through the country rocks, but by heat transferred by permeation outwards from the granite of late-stage H_2O-rich fluid. Laboratory evidence in support of this conclusion will be given in Chapter 14. This probably explains the difference in width in contact aureoles surrounding basic and acid intrusions, because basic magmas usually contain smaller amounts of dissolved volatiles than acid ones.

DYNAMIC METAMORPHIC ROCKS

For this second group of metamorphic rocks, field study also provides clear evidence for the mode of origin. Just as contact metamorphic rocks are found close to igneous intrusions, **dynamic** metamorphic rocks are found close to major fault or thrust planes.

The fault and thrust planes are surfaces in the Earth's crust or upper mantle across which there have been large relative movements, and tectonic studies demonstrate that metamorphism occurred at the same time as movement in cases where dynamic metamorphic rocks have been formed. Faults and thrusts often have a complex history of several phases of movement and dynamic metamorphism may only have occurred during some phases. An example where metamorphism and movement can be closely linked in time by study of the structures in a thrust plane is described in Chapter 6.

There is considerably more variation in the processes of dynamic metamorphism than in those of contact metamorphism. At shallow levels in the Earth's crust, in relatively brittle rocks, there may be intense fracturing leading to fragmentation of rocks on a hand-specimen scale or less. At deeper levels and higher temperatures, the concentration of movement in a thrust or fault plane (or more precisely a narrow zone) may be associated with local recrystallisation of rocks and, in rare extreme cases, with frictional melting. These processes yield distinctive rock types, to be described in this section.

Intense fracturing of rocks in the vicinity of a fault or thrust plane gives rise to a **fault breccia** (Fig. 2.4a). The fine-grained matrix of such a breccia, if examined under a hand-lens, is seen to be a finer-grained microbreccia, with a matrix of 'rock flour', which is too fine-grained for its character to be

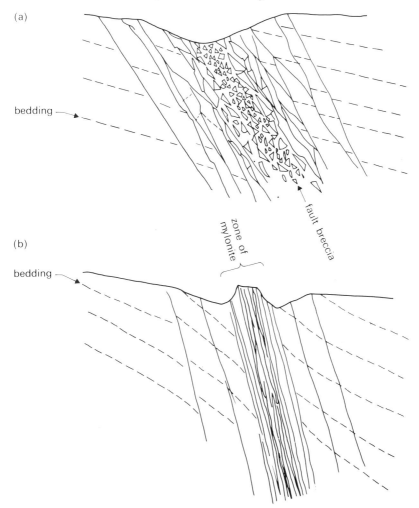

(a)

bedding

fault breccia

zone of mylonite

(b)

bedding

Figure 2.4 Cross sections through fault zones in massive rocks. (a) Fault zone with fault breccia, formed at shallow level in the Earth's crust. (b) Fault zone with mylonite, formed at deeper level.

visible under a hand-lens. The rock flour is susceptible to chemical alteration by percolating ground water and readily alters to fine-grained clayey substance, usually with less altered fragments of rock remaining, which is known as **fault gouge**. The fault gouge or fault breccia with a fault gouge matrix, is relatively easily eroded, which explains why faults are often marked by gullies or parts of stream valleys.

Where the amount of displacement on the fracture is great and the movement occurred at a considerable depth in the Earth so that the walls were pressed together by the weight of the overlying rock, more intense metamorphism of both fragments and matrix occur. The material of the country rock undergoes local recrystallisation, the nature of the resulting dynamic metamorphic rocks depending to some extent on the nature of the country rocks surrounding the fault or thrust. Massive, mechanically strong rocks such as quartzites and, under certain conditions, limestones, give rise to fine-grained, flinty rocks known as **mylonites**. These consist of lens-shaped fragments of country rock in a very fine-grained, flinty matrix. The lens-shaped fragments have a tendency to lie in one plane, and the matrix is often banded in the same direction (Fig. 2.5a). The banding is parallel to the plane of the fault or thrust. The fragments of country rock quite often consist of one crystal only and in that case they are known as **porphyroclasts**. Unlike fault breccia, mylonite is often resistant to erosion, so that a narrow mylonite zone may stand out in the middle of an eroded gully marking a fault zone (Fig. 2.4b).

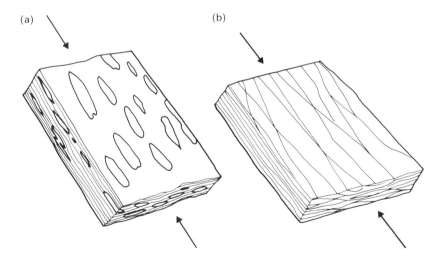

Figure 2.5 Schematic drawings of hand-specimens of (a) mylonite and (b) phyllonite displaying foliation and lineation. The lineation in the mylonite is defined by preferred orientation of the lenticular rock fragments, that in the phyllonite by the intersection of sub-parallel cleavage surfaces.

If the country rock is a shale, slate or schist containing appreciable amounts of phyllosilicate minerals, these may become orientated parallel to the fracture surface giving rise to a tectonic slate, or **phyllonite** (Fig. 2.5b). The orientation of the phyllosilicates parallel to the fracture is less perfect than the parallel orientation seen in regional metamorphic rocks, to be described in the next section. Porphyroclasts are usually not developed. Phyllonites therefore resemble slates and schists (to be described shortly) but split less readily and when split form less flat sheets. The banding of mylonite parallel to the fracture surface, and the tendency of phyllonite to split parallel to the fracture surface, are examples of textures with preferred orientations which are tectonically controlled. Such textures are more often encountered in regional metamorphic rocks. Both the examples described here belong to the class of preferred orientations of a planar form, which are collectively known as **foliations**.

Frequently, in addition to the foliation just described, mylonites and phyllonites display a **lineation**, that is a texture in which there is a preferred orientation of mineral grains or porphyroclasts parallel to one line lying in the plane of foliation. Examples of this are shown in hand-specimens in Figure 2.5. The lineation in mylonites indicates the direction of relative movement of the sides of the fracture during dynamic metamorphism. However, because there have usually been several phases of movement on a fault or thrust and the direction of lineation only gives the direction when the mylonite or phyllonite formed, it must be interpreted with care in discussing the overall movement.

Extreme displacement on fractures in the strong, massive rocks which form mylonites may produce a rock in which all the porphyroclasts have disappeared, leaving a fine-grained, banded rock. This is called an **ultramylonite**. In other unusual cases movement on a fault plane may be so intense that frictional heating of fracture surfaces is sufficient to cause local melting. The melt so formed may intrude the surrounding rocks as small veins, which are diagnostic. The frictional melt solidifies to glass once movement has ceased and the resulting rock type is known as **pseudotachylyte**.

In all the rock types described in this section, the principal process involved in metamorphism is deformation of the country rocks associated with movement on the fault or thrust plane concerned. The responses of rocks to deformation are more varied than their responses to heating in contact metamorphism. They may respond by fracturing, by a considerable variety of types of solid-state intracrystalline and intercrystalline displacement, and in the extreme case by melting due to frictional heating. The particular fashion in which they respond depends upon temperature, pressure, rock composition and several other factors (especially strain rate). A start has been made on determining the range of conditions for certain processes in certain rock types, both experimentally and by a variety of petrological techniques, some new and exciting. But compared with similar efforts at determining the conditions of regional and contact metamorphism (Ch. 13) these investigations are very much at a preliminary stage. For the rock types briefly discussed in this section the identification as dynamic metamorphic rocks is by their association with fault or thrust zones, recognised by geological mapping of rock types, structures, or both.

A different suite of dynamic metamorphic rocks is found in association with circular structures formed by meteorite impact or artificially by underground nuclear explosions. When they are little modified in form by erosion, these circular structures are crater-shaped and they are well-known surface features on the Moon. In the case of impacts on Earth by large meteorites, the kinetic energy of the meteorite is converted directly into very energetic shock waves which travel outwards from the site of the impact diminishing rapidly in intensity the further they go. In underground nuclear explosions, similar shock waves are formed by the very energetic expansion of the material undergoing nuclear fission or fusion, and of surrounding matter which is vapourised by the intense radiation from the nuclear explosion. The passage of the shock waves produces extremely high temperatures and pressures in the rocks for intervals of a few microseconds. This produces metamorphic alteration of the rocks, hence the name **shock metamorphism**. The shock waves produce a shock metamorphic aureole around the crater.

In the outer part of the aureole, shattering of the country rock is seen. The fractures include radial and concentric planar or sub-planar ones, and also non-planar ones. Individual mineral grains are cracked and bent. The intensity of the shock waves means that instead of the fractures being filled with rock flour they are either clean breaks or are filled with glass. Nearer the crater the proportion of glass rises, and close to the site of impact or explosion a breccia-like rock with angular fragments in a matrix of glass is found. This is known as **suevite**. In the innermost part of the crater, specimens of glass free from fragments are found. This material is known as **impactite glass**.

Thus the progressive metamorphic sequence in a shock metamorphic aureole is as follows:

country rock → rock with distorted and cracked mineral grains
↓
rock with glass-filled cracks → suevite → impactite

Like dynamic metamorphic rocks associated with faults and thrusts, natural shock metamorphic rocks are recognised in the field by their association with the characteristic circular structures produced by meteorite impact. Volcanic explosions do not produce shock waves of sufficient intensity to cause shock metamorphism. A number of diagnostic structures indicate that a crater has been formed by shock waves, e.g. overturning of

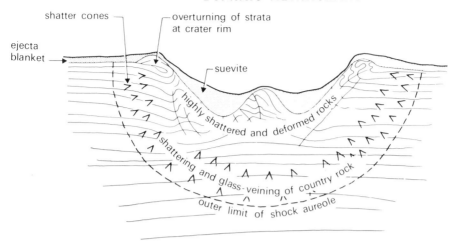

Figure 2.6 Cross section through a meteorite impact crater. Deeper structure very schematic. After Mutch (1972).

strata on the crater rim. These structures are shown in Figure 2.6, which shows a generalised cross section through a crater of about the size of Meteor Crater, Arizona, U.S.A.

On the Moon, many rocks collected from the surface show various stages of shock metamorphism. Because no rocks have been collected from bed rock beneath the lunar **regolith**, these effects cannot be related to progressive shock metamorphic aureoles surrounding particular lunar craters.

REGIONAL METAMORPHIC ROCKS

Regional metamorphic rocks outcrop at the surface over a large part of the Earth. They underlie a comparatively thin layer of sedimentary rocks over another large part (Fig. 2.13). Whereas the principal processes of metamorphism in contact and dynamic metamorphic rocks can be determined by the study of field relationships alone, those in regional metamorphic rocks are more obscure. They are the commonest kind of metamorphic rocks and this section describes some of their characteristic textures and the broad geological settings in which they are found.

Textures in which mineral grains show a tendency to lie in certain directions (i.e. preferred orientation) are characteristic of regional metamorphic rocks (Figs 2.7, 2.8, 2.9). Textures formed by tectonism and metamorphism are often referred to as **metamorphic fabrics**, and the characteristic textures of regional metamorphic rocks as **directional fabrics**. There are two main geometrical classes of regional directional metamorphic fabrics: planar fabrics or **foliations** and linear fabrics or **lineations**. The two fabrics often occur together (Fig. 2.9).

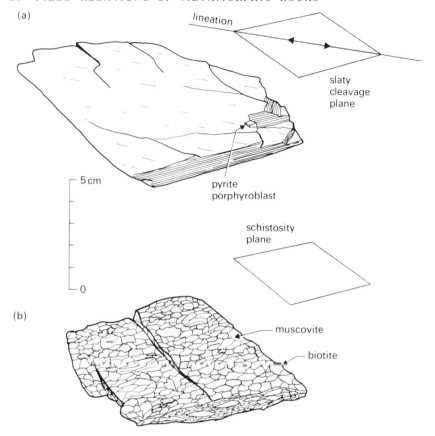

Figure 2.7 Sketches of hand-specimens of (a) slate and (b) schist to illustrate the more regular cleavage direction in slate. Individual muscovite and biotite flakes indicated in schist. Slate specimen from Ballachulish, Argyll; schist from Loch Stack, Sutherland.

Foliation fabrics are those in which the fabric elements, usually the mineral grains, have a planar arrangement. One familiar foliation fabric is **slaty cleavage**. Slates are fine-grained rocks largely composed of phyllosilicate minerals such as muscovite and chlorite. The small tabular crystals of these minerals are arranged so that they are approximately coplanar or, loosely speaking, parallel. Because phyllosilicate minerals have a good cleavage coplanar with the tabular crystals, the rock splits very easily in the direction of the crystals. This property is known as rock cleavage, to distinguish it from the mineral cleavage of the individual mineral grains. If there is no danger of confusion, it is just called **cleavage**. Slaty cleavage is remarkable in that it is so well developed that the rock may be split into very thin planar sheets.

Coarser-grained metamorphic rocks, largely composed of phyllosilicate minerals such as muscovite, chlorite and biotite, also have their individual crystals arranged in a coplanar fabric. Because the crystals are larger and less perfectly coplanar, although rock cleavage is still present, the surfaces are less perfectly flat and the sheets which may be split from the rock are thicker. The cleavage in these rocks is called **schistosity**. Figure 2.7 shows a sketch of a hand-specimen of slate and of schist, showing the different appearances of the two types of cleavage.

Another important type of foliation fabric is a banding of the rock into layers rich in ferromagnesian minerals, such as biotite and hornblende, and layers rich in quartz and feldspars. This banding is seen in the large class of metamorphic rocks known as **gneisses**, and will be referred to as **gneissose banding** in this book, although it is sometimes called gneissosity. Figure 2.8 is a sketch of a hand-specimen of banded gneiss showing the segregation into lighter and darker layers. The scale of the bands can vary considerably, from less than a millimetre to several metres across.

Lineation fabrics are less familiar than foliation fabrics. In rocks with prismatic or acicular mineral grains these may have a parallel arrangement (the word 'parallel' is not being used loosely this time). Figure 2.9a is a sketch of a hand-specimen of granitic gneiss from the Ossola Valley, in the Italian Alps, which shows a lineation fabric in which the parallel elongate crystals are feldspars. Figure 2.9b shows a commoner type of fabric in which a rock with a well-developed foliation fabric (a schistosity) also displays a tendency for prismatic hornblende crystals to run parallel to one direction in the foliation plane. The fabric of the hornblende crystals is a lineation fabric and the rock as a whole displays both foliation and lineation.

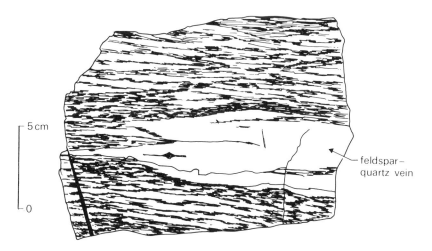

Figure 2.8 Banded feldspar-quartz-epidote gneiss, Myrdal, Norway.

An important distinction should be drawn between **penetrative** and **non-penetrative** fabrics. A penetrative fabric influences every crystal in the rock. Slaty cleavage, in which every phyllosilicate flake tends to be coplanar with the rest, is an example of this. It is not necessary for all the crystals to be precisely oriented parallel to the principal fabric directions; there may be a statistical preferred orientation, as with the hornblende crystals of Figure 2.9b. Non-penetrative fabrics are only developed in parts of the rock and are absent in other parts. Figure 2.10 shows an example of a schist with an early penetrative schistosity (S_1) which has been folded and developed a number of discrete planes of splitting (S_2) parallel to the axial surfaces of microfolds. The S_2 surfaces constitute a non-penetrative cleavage.

Only a few examples of metamorphic fabrics have been discussed here. Others will be described among the examples of regional metamorphic rocks to be discussed in later chapters.

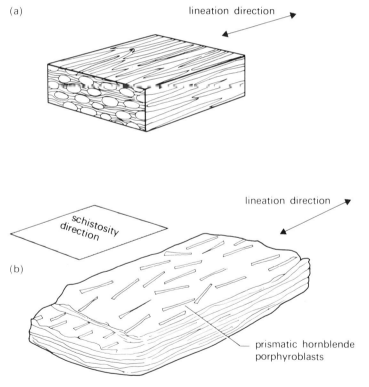

Figure 2.9 (a) Granitic gneiss, Ossola Valley, Italian Alps, a rock with a predominant lineation fabric. (b) Furulund Schist, Sulitjelma, Norway, a rock with combined foliation (schistosity) and lineation.

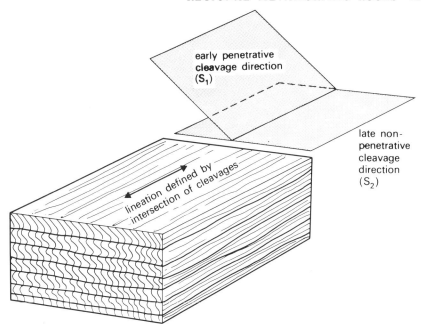

Figure 2.10 Schematic sketch of a schist specimen showing crenulation cleavage, a type of non-penetrative fabric.

It is most important when studying metamorphic fabrics in the field to determine their orientation relative to major and minor geological structures. The systematic study of such relationships is the science of **structural petrology** which lies outside the scope of this book. One crucial relationship between foliation and bedding, however, is so central to much discussion of the evolution of metamorphic rocks that it must be mentioned. When sediments have been folded and have had a **slaty cleavage** imposed on them during the same deformation episode, the cleavage direction is commonly coplanar with the axial surfaces of the folds (Fig. 2.11). This relationship may enable the time of crystallisation of a particular association of metamorphic minerals in the slate layers to be correlated with the deformation episode in which the folds formed. In favourable cases the timing of this episode may be determined by stratigraphical techniques, or the date of metamorphism of the slate may be determined by radiometric methods (Ch. 14). The relative time of formation of the folds, early or late in the deformation history of the area, may be determined from their relationships with other folds, faults or thrusts and from regional comparisons of their directions and styles. The relationship between bedding and cleavage should always be examined carefully in metamorphosed sedimentary rocks. That shown in Figure 2.11 is generally found in low-grade regional

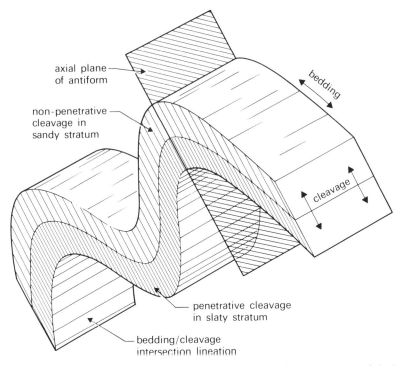

axial plane of antiform

non-penetrative cleavage in sandy stratum

bedding

cleavage

penetrative cleavage in slaty stratum

bedding/cleavage intersection lineation

Figure 2.11 Illustration of the geometrical relationships between axial-plane cleavage and bedding.

metamorphic rocks such as slates. In higher-grade rocks such as schists and gneisses the cleavage is often apparently coplanar with the bedding. A careful search may show that although this is the case in many localities, in a few places cleavage can be found cross-cutting bedding. This reveals the presence of folds that formed during or even before the metamorphism and is a significant result for both the tectonic and metamorphic study of the rocks.

In the regions of multiple deformation in which many regional metamorphic rocks occur, metamorphic foliation may be folded and faulted by post-metamorphic deformation. Deformed metamorphic lineations are especially helpful in the study of such regions. Descriptions of the techniques used in this kind of study may be found in textbooks of structural geology (Ramsay 1967, Turner & Weiss 1963) and will not be given here. It should be emphasised that, although the primary aim in studying a particular region of metamorphic rocks may be structural analysis of this kind, it is nonetheless important to give an accurate account of the chemical and mineralogical compositions of the metamorphic rocks. This may give vital information on conditions during deformation and metamorphism that could help in structural analysis.

In some regions of metamorphic rocks, a variation of grain size and metamorphic minerals occurs which suggests a variation in metamorphic grade like that seen on approaching the contact of an igneous intrusion in a metamorphic aureole. The variation in grade occurs more gradually than in a contact metamorphic aureole, permitting zones of different metamorphic grade distinguished either texturally or mineralogically to be recorded on geological maps.

The best-known sequence of metamorphic zones is that described by Barrow (1912) in the eastern part of the Grampian Highlands of Scotland, but such sequences have been mapped in many parts of the world. Part II of this book describes a number of examples of progressive metamorphic sequences of this kind from different geological settings. The mineralogical and textural changes with increasing grade vary so much depending upon rock composition that to describe any one sequence as 'typical' would be misleading. A few general remarks about progressive regional metamorphism will be made here, but the reader is referred to the later chapters for specific instances to illustrate them.

Progressive regional metamorphic sequences from unmetamorphosed sediments into regional metamorphic rocks are not common, although several have been described (e.g. Chs 10, 11). Regional metamorphism is commonly associated with mountain building processes. As a result, regional metamorphic rocks have almost invariably been folded, thrust and faulted before, during and after metamorphism. It is therefore unusual to be able to pass from unmetamorphosed sediments into regional metamorphic rocks of the same age without crossing a fault or thrust surface at which there is an abrupt change in metamorphic grade. Unmetamorphosed sediments may overlie regional metamorphic rocks of the same composition, but in this case there is usually an unconformity between them representing a considerable interval of geological time.

Most progressive sequences of regional metamorphic rocks are all metamorphic, but of different grades. The higher-grade and lower-grade parts of the sequence can be distinguished by a number of criteria. In general the grain size of the rocks tends to become coarser as the grade increases. At high metamorphic grades the rocks come near to partial melting and veins of granite and pegmatite become common.

The most reliable method for mapping zones of metamorphic grade in progressive regional metamorphic sequences is to record the changes in mineral content in rocks of a particular chemical composition (e.g. **pelitic** rocks) with increasing grade. To do this convincingly it is essential to know all the minerals present in the rocks and to have at least a general idea of their chemical composition. This requires the use of the petrological microscope and therefore will not be described until Part II. However, in some regions of metamorphic rocks the increasing metamorphic grade may be mapped by the incoming of single **index minerals** which may be recognised in the field. This was Barrow's method and he named **metamorphic zones** after diagnostic minerals seen in rocks of pelitic composition in the Dalradian Supergroup. The sequence of zones proposed by Barrow and modified by Tilley (1925), is as follows from low to high metamorphic grade: chlorite, biotite, garnet, staurolite, kyanite, sillimanite. Although the sequence of index minerals is the best indication of the stages of increasing metamorphic grade, there is also a progressive increase in grain size up the sequence, and granite and pegmatite veins become common in the highest-grade sillimanite zone. The mapping of zones by the recognition of a single index mineral in the field, in the Scottish Highlands or elsewhere, depends upon some favourable geological coincidences, for example that the new minerals formed during regional metamorphism grew as **porphyroblasts** (larger crystals in a finer-grained matrix) and that the composition of the pelitic rocks remained constant at different metamorphic grades. Mapping regional metamorphic grade using index minerals without considering the other minerals in the rock is unwise.

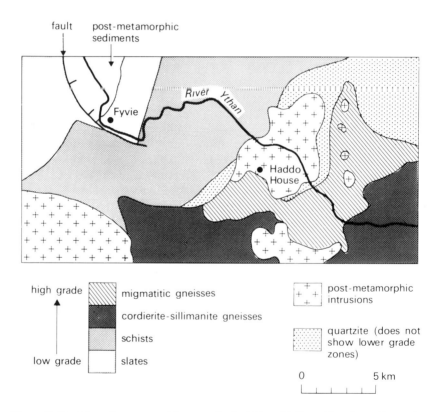

Figure 2.12 Map of metamorphic zones in the Ythan Valley, Aberdeenshire. After Read (1952).

It is possible to map progressive metamorphic zones on textural features rather than mineral changes, although in recent years this has not often been done. Figure 2.12 shows an example from the Ythan Valley, Aberdeenshire, Scotland. This map was published by Read (1952) and records textural changes rather than mineralogical changes in rocks of pelitic composition. The area has since been included in maps of metamorphic zones based on index minerals (Chinner 1966, Porteous 1973).

For the purposes of discussion in Part II of this book, areas of regional metamorphic rocks have been divided into three categories depending on their global tectonic setting: Precambrian shield areas, Phanerozoic orogenic belts, and ocean floors. The metamorphic rocks of Precambrian shield areas are old; those of orogenic belts of many different, but predominantly intermediate, ages; while those of the ocean floors are comparatively young. This age distribution can be explained by plate tectonic theory (Ch. 16). The reason for using a tectonic division of areas of regional metamorphic rocks is that, although some kinds of rocks are found in all three types of area, there are significant differences in predominant rock type and conditions of metamorphism between the three types.

The shield areas are blocks of Precambrian rocks of continental dimensions. According to Read and Watson (1975a, p. 14): 'For the purposes of dealing with the early parts of geological history, the great shields . . . provide the natural divisions on which to base a regional description.' Their advice is followed here. The extent to which the rocks of the shield areas formed by tectonic and metamorphic processes which still operate has been the subject of much controversy. But there appear to be significant differences between the metamorphic rocks of Precambrian shields and those of younger orogenic belts, however they are explained.

The second category of regional metamorphic rocks is restricted to Phanerozoic orogenic belts, although the shield areas also contain the remains of Precambrian orogenic belts. These have some similarities to Phanerozoic belts, but also many differences. Because Phanerozoic belts are more continuous and fossils are preserved in some of their rocks, the history of deposition, deformation, metamorphism and uplift can be determined in more detail than is usually possible in Precambrian belts. As a result the emphasis in the study of their metamorphic rocks is less broad, concentrating on the correlation of tectonic and metamorphic events in one orogenic cycle and relying less upon radiometric age determinations, although these are still extremely valuable.

In this book, regional metamorphic rocks from orogenic belts are discussed in Chapters 8, 9 and 10. Chapters 8 and 9 describe rocks from a Palaeozoic orogenic belt, while Chapter 10 deals with rocks from a Cainozoic orogenic belt. This distinction is made for convenience, not for fundamental petrogenetic reasons.

The study of rocks dredged from the ocean floors, and more recently drilled in the deep sea drilling project, reveals the presence of metamorphic rocks, mostly metamorphosed basic igneous rocks. The rocks of the ocean floors are all comparatively young geologically, the oldest so far discovered being of Jurassic age. The almost total absence of rocks of granitic composition from the ocean floors contrasts with their abundance and importance both in shield areas and in orogenic belts. The composition of ocean floor rocks, and their structure and metamorphism, is well accounted for by their growth by spreading from mid-ocean ridges. This topic is discussed in Chapter 11.

Figure 2.13 shows those parts of the Earth's surface which are underlain by Precambrian shield areas, Phanerozoic orogenic belts, and ocean floors. The areas left without ornament are those where metamorphic rocks are buried under sediments or continental shelf areas where the distribution of the different types is still relatively uncertain.

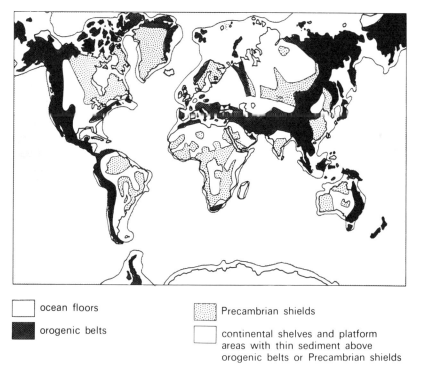

    ocean floors   Precambrian shields

    orogenic belts   continental shelves and platform areas with thin sediment above orogenic belts or Precambrian shields

Figure 2.13 Map of worldwide distribution of tectonic settings of regional metamorphic rocks.

3

Description and identification of metamorphic rocks in the field

There is no generally agreed descriptive classification of the metamorphic rocks, nor are there agreed definitions of such common metamorphic rock types as **schist**, **gneiss** and **amphibolite**. This is confusing for the student and is particularly troublesome in the field, where lack of suitable names can in extreme cases lead to failure to distinguish significantly different bodies of metamorphic rocks.

This chapter aims therefore to provide some guidance in this situation. The names proposed are not intended to be authoritative; they are based on the author's experience and that of his colleagues. It is hoped that they correspond reasonably well with the usage of British geologists, but variations are to be expected around the world, even among English-speaking geologists. There are different schemes in French and German. In general, the student is advised to try to understand and then follow local usages, rather than embark on the re-naming of large areas of metamorphic rocks. If there are no local terms, or if subdivisions are being made of rocks previously placed together under one name, then it is hoped the names outlined here will be helpful.

PRIMARY SEDIMENTARY AND IGNEOUS FEATURES

Many metamorphic rocks retain sufficient of their primary sedimentary or igneous features to be given sedimentary or igneous rock names in the field. Bedding of sediments is often preserved and stratigraphical formations of rocks of different composition may be mapped. Where field study makes it certain that the units of metamorphic rock concerned are primary sedimentary formations, the conventional scheme for lithostratigraphical nomenclature should be employed (Hedberg 1976). Formations, groups, super-groups and so forth may be defined and given appropriate geographical and rock-type names, e.g. Port Askaig Tillite Formation, Mull of Oa Phyllite Group, Dalradian Supergroup. Sedimentary, igneous or metamorphic rock names may be used.

Igneous rocks may also retain sufficient of their primary features to be identified as lava flows, bedded pyroclastic rocks, minor intrusions or major intrusions. Lava flows and pyroclastic rocks may be named according to the lithostratigraphical conventions mentioned above. Intrusive igneous rocks usually have rock names distinct from their extrusive equivalents and it may be appropriate to use these to distinguish intrusive from extrusive rocks, even though both are metamorphosed.

If it is necessary to emphasise that a particular rock type has undergone metamorphism, this may be done in two alternative ways. The prefix 'meta-' may be put before the appropriate igneous or sedimentary rock name, e.g. Tayvallich Metabasalt, Knoxville Metagreywacke, or a metamorphic rock name may be used in preference to an igneous or sedimentary one, e.g. Furulund Schist. Often the context makes it clear that metamorphic rocks are being described, making it tedious to repeat the prefix 'meta-'. Therefore unmodified sedimentary and igneous rock names are often used for metamorphosed or partly metamorphosed rocks, e.g. Loch Tay Limestone, Sulitjelma Gabbro. It is pedantic to insist that such names be modified merely to preserve a uniform convention for naming metamorphic rocks.

In many areas, however, primary igneous and sedimentary features have been completely destroyed by metamorphism. In others it is not certain whether the boundaries between different compositional types of metamorphic rocks represent sedimentary bedding or not. In such cases, the different units of metamorphic rock should *always* be given metamorphic names, e.g. Man of War Gneiss, Landewednack Hornblende Schist. The use of stratigraphical terms such as 'group' should be avoided, although unfortunately this is not always done in scientific publications. Several metamorphic rock units may be referred to collectively by the descriptive name 'complex', e.g. the Lizard Complex (Hedberg 1976, p. 34). This is preferable to the older practice of adding the suffix '-ian' to a geographical name, i.e. 'the Malvern Complex' is preferable to 'Malvernian' because the use of this suffix inevitably leads to confusion with stratigraphical stage and system names. Unfortunately, some names such as 'Lewisian' are too well established to be replaced, but it may be possible to prevent others, such as 'Moinian' from becoming similarly entrenched.

IDENTIFICATION OF THE MINERAL ASSEMBLAGE

Wherever possible, the minerals in a metamorphic rock should be identified in the field. They should be listed in diminishing order of abundance, e.g. biotite, muscovite, garnet, quartz. Such a list is not necessarily a **mineral assemblage** list, because some minerals may be too fine-grained to identify and also because a list of the minerals in a metamorphic rock is not always a metamorphic assemblage (Ch. 4). But field lists of this sort are often useful in identifying significant changes in rock composition and metamorphic grade (Chs 5, 8, 9, 10). If such changes can be identified in the field rather than in the laboratory, much time and effort can be saved.

The techniques for identifying minerals with the aid of a hand-lens, penknife and acid bottle will be found in any elementary textbook of mineralogy (e.g. Battey 1972) and will not be repeated here. The student should apply the techniques to grains smaller than the large specimens encountered in mineralogy classes, and it will be particularly helpful if hand-specimens of metamorphic rocks are studied for which thin sections are also available. The trick when doing this is to prepare a hand-specimen assemblage list before examining the thin section. It is worthwhile taking considerable time over mineral identification at the beginning of a field study, in order to avoid having to put right incorrect observations later on. One common mistake in hand-specimen description of labelled rocks from teaching or museum collections is to imagine that one has identified the minerals which the name indicates should be present. For example, hornblende and plagioclase feldspar might be identified *after* reading the label 'amphibolite' on a specimen. It is much easier to give the name 'hornblende' to the dark mineral in a rock so labelled than it is to identify the mineral without knowing the rock name. The student should guard against this by taking care to do some of his study on unlabelled specimens!

METAMORPHIC FABRIC

Certain textural features of metamorphic rocks play a key role in their description and identification. The term metamorphic fabric, used to describe such textural features, has already been introduced in Chapter 2. Preferred orientation fabrics, such as foliation and lineation, are especially important in this regard. Cleavage and schistosity are foliation fabrics which are essential to slate and schist respectively. It is possible to have metamorphic rock types of the same composition as slate and schist, but lacking the slaty cleavage or schistosity. If these occur in contact aureoles, they are called hornfelses, if in regional metamorphic areas they are called **argillites** at low metamorphic grades and **granofelses** at high metamorphic grades. Another important type of foliation fabric for the identification of metamorphic rock types is the segregation into light- and dark-coloured bands known as gneissose banding. This type of layering is called 'banding' rather than 'layering' to avoid any implication that it is sedimentary stratal layering (i.e. bedding). It is a non-penetrative fabric and is common in high-grade regional metamorphic rocks. It is an essential fabric for some types of gneiss.

The directional fabrics of metamorphic rocks and also such non-directional textural features as grain size or **porphyroblastic texture**, should be as carefully recorded as the mineral assemblage for the identification of metamorphic rocks. Often, textural or fabric features are useful for the identification of a particular metamorphic rock unit. For example the Furulund Schist formation of the Sulitjelma region of Norway (Ch. 8) is characterised by porphyroblastic hornblende needles lying in the schistosity plane (Fig. 2.9b) and one rock type within the Sulitjelma Amphibolite Group is characterised by relict porphyritic texture in metamorphosed dolerites.

METAMORPHIC ROCK NAMES

Two features of metamorphic rocks have now been discussed which may be used in assigning them names: rock composition (reflected in the mineral assemblage) and metamorphic fabric. For some rock types (e.g. **marble**) the first is more important, for others (e.g. **slate**) the second. The two features are not independent; a rock can only have a slaty cleavage if it contains a high proportion of phyllosilicate minerals. One aspect of metamorphic fabric is independent of composition, however, and that is the grain size. It has already been shown that this tends to increase with increasing metamorphic grade in both contact and regional metamorphic rocks.

Table 3.1 divides the metamorphic rocks into broad compositional classes representing the commoner types of sedimentary and igneous rocks. The classes are subdivided into three divisions on the basis of grain size: fine-grained, with grains less than 0·1 mm in diameter; medium-grained, with grains from 0·1 mm to 1·0 mm in diameter; and coarse-

(a) diameter of spots
   0·1 mm

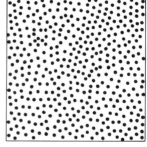

(b) diameter of spots
   1·0 mm

Figure 3.1 Patterns of dots of (a) 0·1 mm diameter and (b) 1·0 mm diameter.

Table 3.1 A proposed scheme of classification for metamorphic rocks in the field.

Primary igneous or sedimentary rock composition	Main mineral groups present	Fine-grained < 0·1 mm	Medium-grained 0·1 mm – 1·0 mm	Coarse-grained > 1·0 mm homogeneous	banded
shale 'pelitic rocks'	phyllosilicates, quartz	slate	phyllite	schist	gneiss
sandstone 'psammitic rocks'	quartz	← quartzite →			banded quartzite quartzitic gneiss
limestone	calcite, dolomite	← marble →			banded marble
marl	phyllosilicates, calcite, dolomite	calcareous slate	calc-phyllite	calc-schist	calcareous gneiss
sandy shale 'semi-pelitic rocks'	quartz, phyllosilicates	quartzose slate, cleaved quartzite	semi-pelitic phyllite	semi-pelitic schist	gneiss
basalt, gabbro	amphiboles, plagioclase feldspar	greenschist greenstone	amphibolite	amphibolite	hornblende gneiss pyroxene gneiss
rhyolite, granite	K-feldspar, quartz, phyllosilicates	halleflinta	← granitic gneiss →		
dunite, pyroxenite, peridotite	serpentine, talc, Mg amphiboles	serpentinite	serpentinite ← talc schist → soapstone		ultramafic gneiss

grained, with a grain size greater than $1 \cdot 0$ mm. To help with the recognition of grain size, Figure 3.1 shows a pattern of dots with mean diameters of $0 \cdot 1$ mm and 1 mm. The tendency for high-grade metamorphic rocks to develop gneissose banding has already been discussed. The coarse-grained division of metamorphic rocks in Table 3.1 has been further subdivided into those showing gneissose banding and those remaining texturally homogeneous.

The rock composition names of Table 3.1 require some further explanation. It is common practice among geologists studying metamorphic rocks to use the adjective **pelitic** to refer to rocks having the chemical composition of shales and slates, and **psammitic** to refer to rocks having the composition of sandstones. Quite often the nouns 'pelite' and 'psammite' are used as rock names (e.g. Miyashiro 1973), or the prefix 'meta-' is added to make 'metapelite' and 'metapsammite'. The names come from an obsolete scheme of classification of sedimentary rocks and some geologists object to them because they are being used in a very different sense from their original meaning. They have the merit that they provide a convenient name for broad compositional categories of rocks, which is helpful in discussing metamorphic rocks. 'Pelitic' and 'psammitic' are used as adjectives in this book; the nouns 'pelite' and 'psammite' are not used. It is the author's view that they should be confined to names for general compositional classes of rocks and not used as field names for individual rocks or rock units, for which the traditional names schist, quartzite or slate are perfectly adequate. An exception, however, is made for the name **semi-pelite** which is a useful field name for quartzose schists and schistose quartzites, which are common for example in the Moine Nappe Complex of northwestern Scotland, and for which no other name is readily available. The old term **grit** used for these rocks has been applied to such a variety of sedimentary and metamorphic rocks that it has become too ambiguous to use.

The name **phyllite** is applied in Table 3.1 to rocks of pelitic composition, with a cleavage and a grain size intermediate between slate and schist. In some progressive metamorphic sequences it is helpful to have an additional name for such rocks (Fig. 2.12), although many geologists simply describe such rocks as fine-grained schists. The distinction between slate and phyllite, if the name is used, is a rather clear one. In slate the individual phyllosilicate crystals are too small to be seen under a hand-lens, whereas in phyllite they can be seen. The boundary between phyllite and schist at a grain size of 1 mm, proposed in Table 3.1, is arbitrary and the reader should not be surprised to find rocks quite frequently described as schists which have finer grain sizes than 1 mm. In some sections of Table 3.1 several rock names are given. Quartzose slate and cleaved quartzite are alternative names, and the reader may use one or the other depending on whether quartz or phyllosilicates predominate in the rock. **Greenschist** and **green-**

stone differ texturally: greenschist has a schistosity, greenstone has not and is therefore a massive rock. Although **serpentinite**, **talc schist** and **soapstone** are all metamorphosed ultrabasic igneous rocks, serpentinite is predominantly composed of minerals of the serpentine group, soapstone and talc schist of talc. Soapstone and talc schist differ in their texture, the former being massive, the latter having a schistosity.

The metamorphic rock names in Table 3.1 may be prefixed by one or more mineral names if it is necessary to emphasise the presence of certain minerals (e.g. forsterite marble, andalusite slate, garnet-mica schist). The presence of a mineral name in such a prefix does not imply that the mineral predominates in the rock, nor does the order of mineral names, if there are more than one, imply their order of abundance. Textural features of a metamorphic rock may also be emphasised by appropriate adjectives (e.g. **augen** gneiss, spotted slate, **saccharoidal** marble) but it is tautologous to prefix an adjective describing an essential textural feature of the rock (e.g. 'cleaved slate'). 'Banded gneiss' is sometimes used because, as the table shows, some gneisses are rocks of granitic composition with a foliation fabric which is not the segregation into light and dark bands described above as gneissose banding; it may therefore sometimes be desirable to distinguish these 'granitic gneisses' from 'banded gneisses' of the more familiar type. Primary textural features may also be emphasised in prefixes (e.g. porphyritic amphibolite, current-bedded quartzite).

There are also special names for some metamorphic rocks with particular modes of origin. Examples which have already been given are **phyllonite** and **mylonite** for dynamic metamorphic rocks, and **hornfels** for contact metamorphic rocks; others will be encountered throughout this book. In appropriate cases these should be used in preference to the general names described in this chapter.

The reader will appreciate from this account that a large degree of latitude is permissible in naming metamorphic rocks in the field. He should try to be as systematic as he can and always define carefully the rock names he gives. It is to be hoped that if this practice is followed, metamorphic rock names will gradually become more uniform until they become subject to an international agreement, like that currently under discussion for igneous rocks. This happy state of affairs is a long way off at present.

Part II

Metamorphic rocks under the microscope

A NOTE ON THE THIN SECTION DRAWINGS

The thin section drawings were made freehand directly from the microscope, not using photomicrographs or an aid such as the camera lucida. I have therefore done nothing that the reader of this book cannot do for himself if he has thin sections and a petrological microscope. They all represent the thin sections viewed in plane polarised light, which I have found is clearest in the great majority of cases. I have also followed the example of Harker (1932) in drawing actual portions of specific thin sections rather than idealisations of features from several parts of a section or even from more than one section. I have drawn the thin sections rather than presenting photomicrographs in order to indicate the mineral species of each grain in each rock by means of suitable ornaments. In photomicrographs, although rock textures are admirably displayed, many grains cannot be identified mineralogically. I have tried to keep the ornaments consistent throughout, and the more common ones are shown in the Figure below. Minerals not shown below are labelled when they occur. Occasionally ornaments different from those in the Figure have been used for clarity, and in these cases the change is indicated by a label.

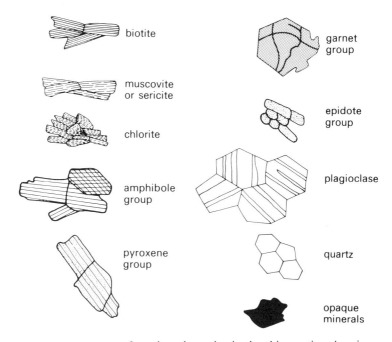

Some common ornaments for mineral species in the thin section drawings.

4

The chemical and mineralogical compositions of metamorphic rocks

INTRODUCTION

In this book, the main object of the study of metamorphic rocks is taken to be the discovery of the conditions of metamorphism. The petrological microscope helps with this task in two ways: the minerals in the rock may be identified and the textures studied in much closer detail than is possible in hand-specimen. With a view to establishing which minerals coexisted during metamorphism, the identification of minerals is emphasised rather more than the description of textures. This is because, in the writer's opinion, current petrogenetic theories linking the coexisting minerals in metamorphic rocks to the conditions of metamorphism provide more consistent and comprehensive results about conditions and processes inside the Earth than do interpretations of metamorphic textures. A comprehensive account of metamorphic textures would have to go beyond petrology and include a good deal of structural geology. The interested reader is referred to the textbooks of Spry (1969) and Vernon (1976) on metamorphic textures, and the recently published textbook of Hobbs *et al.* (1976) on structural geology, for more information on these subjects.

In this chapter, a few observations on the use of the petrological microscope will be followed by discussion of the methods used to describe the chemical composition of metamorphic rocks and to establish lists of minerals coexisting in equilibrium (which raises the question of the meaning of 'equilibrium' in this context). Finally, there will be an outline of methods for representing chemical and mineral compositions of metamorphic rocks on different types of diagram.

THE PETROLOGICAL MICROSCOPE

The petrological microscope is a fairly complicated piece of apparatus and its use demands a good deal of knowledge and skill. It is assumed that the reader is already familiar with the operation of the instrument. Those who are not are referred to Hartshorne & Stuart (1950), to the section describing the instrument in Battey (1972) or to comparable works. Some aspects of recent developments in the microscope itself and in the preparation of thin sections will be mentioned here, because they change microscopic techniques compared with the traditional methods described in the works mentioned above.

Figure 4.1 illustrates a typical modern petrological microscope using transmitted light, the Swift model MP.120. Such modern instruments are

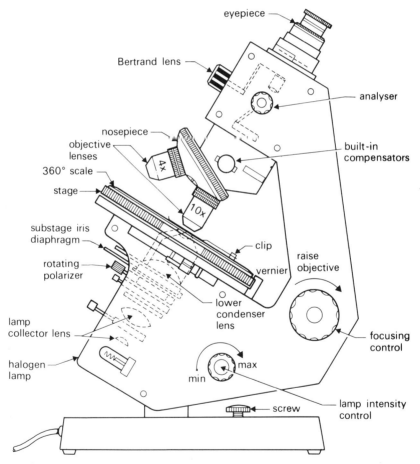

Figure 4.1 The Swift Model MP.120 Polarising Microscope. Courtesy of J. H. Bassett, James Swift & Sons.

easier and quicker to set up than the traditional type. The greatest improvement has been the incorporation of the light source into the instrument, which means that its intensity can be adjusted easily and lengthy readjustments need not be made every time the instrument is set up. Accessory equipment, such as the sensitive tint plate, is built in. The improvement in convenience and the increased number of built-in facilities have resulted in more differences in the layout of controls between different models of microscope than there used to be and it is therefore more important that the student takes time to become familiar with the particular instrument he is using. One remarkable achievement in recent years has been the mass production of a petrological version of the McArthur microscope for the Earth Science courses of the Open University. Although this has neither the range of facilities nor the high optical quality of the Swift microscope shown in Figure 4.1, it is very compact and remarkably cheap. The petrographic observations in this book could have been made with a McArthur microscope, except for the high-magnification studies such as that of fibrolitic sillimanite (Fig. 9.3).

The preparation of thin sections, usually carried out by a skilled laboratory technician, has also been changed in recent years. It is now possible for machinery to perform the delicate operation of grinding down the slice of rock to the exact thickness required (30 μm in most cases). This has been made possible by the development of cold-setting resins such as cyanoacrylate adhesives for mounting the rock slice onto a glass slide. These stronger mounting media prevent the rock becoming unstuck from the slide during mechanical grinding, which is more rapid and less gentle than hand-grinding. Although mounting media with a refractive index of 1·53, like Canada Balsam, are now generally available, thin sections may have been prepared using media with other refractive indices, such as 'Araldite'. The student should bear this possibility in mind. Also, the old process of hand-grinding meant that every thin section was examined under a microscope during the final stages of grinding by an experienced technician who could recognise and eliminate artifacts such as bubbles, fragments of grinding powder, cracks or holes in the rock slices. This is no longer necessarily the case and the student should be more on his guard. The great gain from mechanisation is that the technician, who is as essential as ever to operate the complicated expensive machinery, can make many more thin sections per day. Therefore it is possible for an increased number of thin sections to be made to match the increased number of chemical analyses of rocks which modern apparatus and techniques can perform, and it is still possible for every rock which is chemically analysed also to be examined in detail under the petrological microscope.

The petrological microscope enables all the transparent minerals in a rock to be identified, except in certain rare awkward cases. Opaque minerals may similarly be identified using a reflected-light microscope on a polished surface of the rock. It is possible to polish the surface of a thin section, and to use a combined transmitted-light and reflected-light microscope to make a complete petrographic description of the rock. This has not been done in this book, where opaque minerals are usually not identified, being described simply as 'opaques'. Combined reflected-light and transmitted-light microscopes are not yet in general use by students, nor are polished thin sections widely available.

The petrological microscope also provides a great deal of information about the sizes and mutual relationships of the mineral grains in the rock. Petrologists traditionally call this the **texture** of the rock. Much insight into the interpretation of metamorphic rock textures has come from comparisons with metallurgy and materials science. Scientists working in these fields use the word 'texture' in a different sense. Therefore many specialist publications on this subject, such as Vernon (1976), use the name **microstructure** instead. This usage has not yet become common and 'texture' is used in its traditional sense in this book.

METAMORPHIC MINERALS

Unlike its companion work on igneous rocks, this book does not include a section specifically describing metamorphic minerals. This is because many minerals which are common in metamorphic rocks are also common in igneous ones and it is assumed that the reader is familiar with them. Features of the mineralogy of minerals peculiar to metamorphic rocks will be described in the appropriate places in the text. For the data needed to help with the identification of minerals using the petrological microscope the reader is referred to books on optical mineralogy (e.g. Kerr 1959, Heinrich 1965). Deer, Howie and Zussman (1966) is particularly recommended for the information it gives on the optical properties of minerals and on their chemical compositions, which will often be discussed in this book.

ROCK COMPOSITION

The Earth has an oxygen-rich atmosphere and the rocks of its outer layers, the crust and upper mantle, are also relatively oxygen-rich. Oxygen is the most abundant element in both igneous and sedimentary rocks. It has therefore become conventional to express the composition of rocks as a list of weight percentages of oxides. This tradition began in the days when rock analysis was carried out by dissolving the rock and separating the different chemical elements from solution by a refinement of the 'group precipitation' procedure familiar to chemistry students. The separated precipitates were heated to constant weight in air, so that in many cases an oxide residue was weighed directly and expressed in the analysis as a proportion of the weight of the sample. Although most rock analyses are not now made in this way the old form of presentation is retained by petrologists, to the amusement of chemists!

Ten major oxides are usually quoted, which make up as much as 99% by weight of most silicate rocks. In the usual order in which they are given they are: SiO_2, Al_2O_3, Fe_2O_3, FeO, MgO, CaO, Na_2O, K_2O, 'H_2O+' and 'H_2O-'. Note that the separate quotation of Fe^{+2} and Fe^{+3} oxides indicates that in many rocks the iron is not completely oxidised. The expressions 'H_2O+' and 'H_2O-' are especially idiosyncratic. Most rocks contain water in pore spaces or adhering to the surfaces of mineral grains (analysis of rocks usually begins from a crushed powder). The minerals may also contain water locked in the structure and hydroxyl groups combined with the cations of the mineral structure. Experience has shown that the water adhering to or adsorbed upon mineral grains and that loosely held in cracks is fairly reliably determined by measuring the weight driven off when the powder is held for several hours at a temperature of 105 °C in an oven. This is quoted as 'H_2O-'. The more tightly bound water and the hydroxyl ions are driven off by much stronger heating and are quoted as 'H_2O+'.

A number of minor elements are also quoted as oxide weight percentages. The selection of these depends on the composition of the rock being studied. Some non-metals, such as fluorine and chlorine, thought to be present in the rock as anions F^-, Cl^-, are not quoted as oxides but in element weight percentages. Non-metals often quoted in this way include F, Cl, and S. Minor elements commonly quoted as oxides include TiO_2, Cr_2O_3, MnO, BaO, P_2O_5, CO_2 and SO_3. In some kinds of rocks these may be major oxides. Finally, the total of the weight percentages of the oxides and elements quoted is given. This is intended to indicate the reliability of the analysis, totals between 99·7% and 100·5% being regarded as acceptable. (Examples of rock compositions given in this weight percentage of oxides form are to be found at the end of Chapters 5, 6, 7, 8 and 9, in the calculation problems.)

This form of presentation of rock analyses is not a very useful one for petrological studies on metamorphic rocks. It is the relative *numbers* of atoms of the different elements in the rock which are of interest. Because of the oxidised condition of rocks, the relative numbers of molecules of the different oxides in the analyses are often compared. To obtain these the weight percentages of the oxides are divided by their molecular weights. A list of molecular weights is given in the Appendix.

In order to make the distinction between weight percentages and molecular proportions clear, square brackets will be put round the oxide formula when the molecular proportion is under discussion (e.g. $[SiO_2]$, $[Fe_2O_3]$, $[TiO_2]$). When weight percentages are being referred to, the numbers will be left without brackets.

MINERAL COMPOSITIONS

Mineral analyses are commonly quoted in two forms: as an oxide weight percentage list as for rocks, and as molecular proportions, usually expressed as a chemical formula. Deer, Howie and Zussman (1966) give mineral compositions in both forms and the method of calculation of the formula from the chemical analysis is given in Appendix 1, of their book.

Because minerals are chemical compounds, the relative proportions of elements and therefore oxides in their analyses are usually fixed. For example $[CaO]/[SiO_2]$ in diopside is $1/2$, in tremolite $1/4$. This is not true of all oxides in all minerals. Where one metallic ion may substitute for another in the same position in the crystal structure of a mineral, a phenomenon known as **diadochy** or **solid solution**, the ratio of these two may have any value (e.g. $[FeO]/[MgO]$ in diopside-hedenbergite) or may have any value within a fixed range (e.g. $[Al_2O_3]/[SiO_2]$ in plagioclase feldspars, for which the range is from $1/3$ in albite to $1/1$ in anorthite). The molecular ratios of the sums of the substituting oxides remain fixed relative to other oxides in the mineral. For example $[FeO+MgO]/[SiO_2]$ in diopside-hedenbergite is $1/2$.

A mineral such as a clinopyroxene in the diopside-hedenbergite series constitutes one member of the suite of minerals in a metamorphic rock. As will be shown in a later section, this may be expressed by saying that it is a single chemical phase. Its chemical composition may however be expressed as proportions of the two end-member compositions (e.g. 68% $CaMgSi_2O_6$, 32% $CaFeSi_2O_6$). The $CaMgSi_2O_6$ and $CaFeSi_2O_6$ are the two **components** of the clinopyroxene phase.

Much of this is probably already familiar to the reader from the study of mineralogy, but it has been repeated here because it is crucial to the understanding of what follows.

EQUILIBRIUM

Many metamorphic rocks have a comparatively simple mineral composition, consisting of one, two, three, four or five major mineral species. This does not include those accessory minerals which take up virtually all of a particular minor chemical element in the rock, for example apatite containing phosphorus or zircon containing zirconium. This mineralogical simplicity is thought to be due to the attainment by the rock of a state of thermodynamic equilibrium during metamorphic recrystallisation. Provided that the rock remains at a high temperature and pressure in the solid state for a sufficiently long time, recrystallisation will go on until the rock has been entirely reconstituted into a new association of minerals stable under these conditions. The new minerals will be those possessing the lowest chemical potential energy under the conditions of metamorphism. This is not a rigorous definition of equilibrium, which should be given in thermodynamic equations. The interested student will find these in Turner (1968, pp. 63–75).

Figure 4.2 illustrates diagrammatically one type of change from a hypothetical sedimentary rock to the equivalent metamorphic rock, involving the progressive attainment of thermodynamic equilibrium. The rock illustrated has a simple chemical composition and at all stages contains only two or three minerals. Before metamorphism, as a sedimentary rock, it consists of 25% of mineral A and 75% of mineral B. This is shown in Figure 4.2a. Figures 4.2b and 4.2c show the rock at the elevated temperature and

(a)
original
sediment

(b)
rock goes to
higher
temperature,
pressure;
reaction
incomplete

(c)
rock continues
at high
temperature,
pressure;
reaction
complete

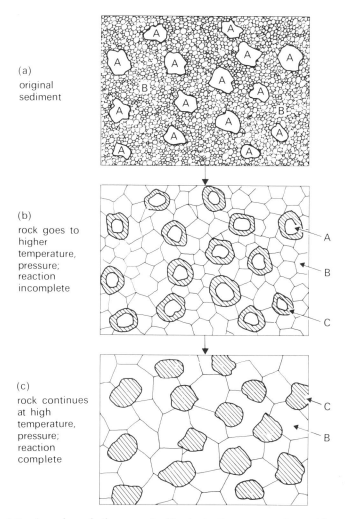

Figure 4.2 A series of diagrams to illustrate the progressive achievement of equilibrium during metamorphism of a hypothetical two-mineral rock.

pressure of metamorphism. Under these conditions the minerals A and B do not exist stably together. A chemical reaction takes place between them, which may be written as follows:

$$A + B = C$$

The mineral C has a lower chemical potential energy at the temperature and pressure of metamorphism than the equivalent amounts of A and B. B on its own, however, remains stable under these conditions. The chemical reaction will continue until all mineral A is used up, because there is an excess of mineral B in the rock. The exact proportions by weight of mineral A and mineral B used up to make mineral C in the chemical reaction can be determined if the chemical equation for the reaction is known. The relative volumes and the total volume change can then be worked out if the densities of the minerals are known.

The rock as shown in Figure 4.2b is in a state of disequilibrium, because all mineral A has not yet reacted with mineral B. Figure 4.2c shows the rock in a state of equilibrium *for the conditions of metamorphism*. Under these conditions the minerals B and C are an equilibrium mineral assemblage. Equilibrium mineral assemblages of this kind will be referred to simply as **mineral assemblages** in this book. Many petrologists use the name mineral **paragenesis** instead. Mineral assemblages will be written as lists of mineral names separated by plus signs, thus:

diopside + anorthite + grossular garnet + quartz

Therefore the hypothetical simple rock of Figure 4.2c shows the mineral assemblage B + C.

Rocks which are more complex chemically and mineralogically than this simple hypothetical example have a larger number of minerals in their equilibrium mineral assemblages. In a later section it will be shown that there is a simple relationship between the chemical composition of the rock and the number of minerals in its equilibrium mineral assemblage. There is no simple way to recognise whether a metamorphic rock has achieved equilibrium. Certain textures are often associated with equilibrium (e.g. **granoblastic texture**) but rocks without these may also have equilibrium mineral assemblages. The author assumes that metamorphic rocks have equilibrium mineral assemblages unless there is definite evidence to the contrary. This is a sweeping assumption and in some rocks it can be shown to lead to wrong results because the 'assemblage list' based on this assumption contradicts well-established relationships between metamorphic minerals discovered by laboratory experiment (Ch. 13). However, an occasional error is outweighed by the gain in understanding of the conditions of metamorphism obtained from the study of mineral assemblages.

The textural relationships which must be demonstrated if a rock is to be regarded as having an equilibrium mineral assemblage are as follows:

(1)   Each mineral in the assemblage list must have a boundary somewhere in the rock with all the other members.
(2)   The texture must be of a type thought to have formed by metamorphic recrystallisation, not by fragmentation during dynamic metamorphism or igneous crystallisation from a melt.
(3)   The minerals must not show compositional zoning (Ch. 15).
(4)   The minerals must not show obvious replacement textures such as reaction rims or alteration along cracks.

The application of these criteria requires experience. The examples described in this book are designed to give the student such experience at second hand, but it is obviously better to acquire experience by direct study. A student who has done this will be able to compare the examples described here with the rocks he has studied and develop his own judgement in the matter. The more rocks studied the better.

THE PHASE RULE

Since Goldschmidt (1911) examined the hornfelses surrounding igneous intrusions in the Oslo region, Norway, it has been recognised that there is a simple relationship between rock compositions and mineral assemblages for metamorphic rocks which have attained thermodynamic equilibrium. This relationship is an application to metamorphic rocks of a simple rule of physical chemistry, the **Phase Rule**. The Phase Rule is expressed as the equation:

$$P + F = C + 2 \qquad (1)$$

P, F and C are simple whole numbers (integers). P is the number of **phases** in the system, that is to say the number of physically distinct kinds of substance which may be distinguished. F is the number of modes of variation or **degrees of freedom** of the system. C is the number of chemical **components** in the system. The rule applies to any chemical system in equilibrium. It may be applied either to a rock undergoing metamorphism or to an experimental system under observation in the laboratory.

These definitions, and the Phase Rule itself will be explained with the aid of Figure 4.3, which illustrates a simple series of experiments with a sealed vessel containing pure H_2O only.

The temperature in the vessel may be changed by heating or cooling from outside, and the pressure by means of a piston at one end. Figure 4.3 shows the state of the experimental system (everything inside the vessel) under

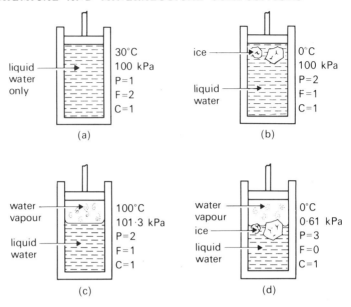

Figure 4.3 Experiments in the one-component system H_2O.

different conditions during the experiment. Three different phases appear during the experiments, liquid water, ice and water vapour. All have the same composition, H_2O, and H_2O is the *only* substance inside the vessel during the experiment. At all stages of the experiment therefore, there is only one chemical component to be considered, H_2O. On the right of Figure 4.3 are the values of P, F and C in equation (1). At all stages $C = 1$.

Figure 4.3a shows the system at a temperature of 30 °C and under a pressure of 100 kPa (approximately 1 atmosphere). The system contains only one phase, liquid water. Therefore $P = 1$. It follows from equation (1) that $F = 2$, i.e. the system has two degrees of freedom. What does this mean? It is possible to change *either* the temperature *or* the pressure on the system as shown in Figure 4.3a by a small amount without changing the state of the system. If the temperature is increased to 31 °C or decreased to 29 °C, the vessel will still only contain water. If the pressure is increased to 110 kPa or decreased to 90 kPa, the vessel will still only contain water. The temperature and pressure may be varied independently. The system is said to be **divariant**.

In Figure 4.3b, the system contains two phases, water and ice. $P = 2$ and so from equation (1) $F = 1$. The system is said to be **univariant**. In this case a small change in either temperature or pressure would cause a fundamental change in the system, reducing the number of phases present from two to one. An increase of temperature to 1 °C, keeping the pressure constant at 100 kPa, would cause all the ice to melt. A decrease of temperature to -1 °C would cause all the water to freeze. An increase of pressure to 110 kPa, if temperature is kept constant at 0 °C, would cause all the ice to melt, a decrease of pressure to 90 kPa would cause all the water to freeze. Only if the *ratio* of temperature to pressure were kept constant would the system continue to contain two phases, ice and water.

In Figure 4.3c, the system also contains two phases, water and water vapour (or steam). In this case an increase in temperature to 101 °C will cause all the water to change to steam, if the pressure is kept at 101·3kPa, while a decrease in temperature to 99 °C will cause all the steam to condense to water. An increase in pressure to 102·3 kPa will cause all the steam to condense to water, if the temperature is kept at 100 °C, and a decrease of pressure to 100·3 kPa will cause all the water to change to steam. At 99 °C water and steam will coexist, so the system remains a two-phase one, at a pressure of 97·9 kPa, and at 101°C a two-phase system will be preserved at 105·0 kPa. For any temperature, the pressure is fixed, and vice versa. This is strictly what is meant by describing the system as **univariant**.

Figure 4.4 shows the univariant relationships between water and ice and water and water vapour on a temperature–pressure graph. It can be seen that the divariant states of the system, when there is just one phase present, are represented as areas on the graph. The univariant states are represented by curves separating the divariant fields, and the particular examples of univariant states of the system shown in Figure 4.3 are indicated on the diagram. Lastly there is an **invariant triple point** at which ice, water and vapour all coexist. The temperature and pressure for this state of the system are unique, and any change in them will cause one or more phases to disappear. This state of the system is shown in Figure 4.3d. Figure 4.4 is called a **phase diagram**. It is a simple diagram because the system described is a simple system with only one component. One-component rock systems do exist, the most important being the SiO_2 system.

However, the discussion of rock compositions and mineral assemblages earlier in this chapter will make it clear that most metamorphic rocks have several components. From equation (1) it will be seen that as the number of components increases, the number of phases also increases. If the rock was in equilibrium at the time of metamorphism, the minerals in the rock can be regarded as phases. Thus the Phase Rule can link the number of minerals in the assemblage list with the number of chemical components in the rock. This explains the fact, mentioned at the beginning of the chapter, that the number of minerals in most metamorphic rocks is quite small.

How many degrees of freedom are likely to be present in rocks undergoing metamorphism? Temperature and pressure are likely to be deter-

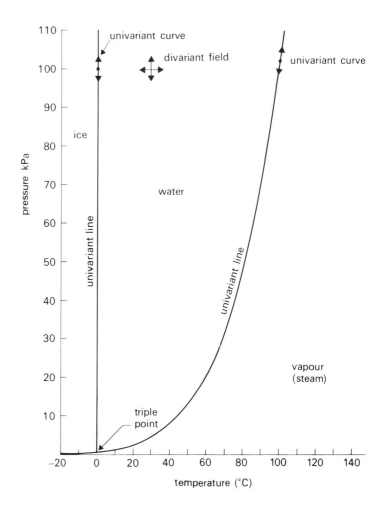

Figure 4.4 Phase diagram of the one-component system H_2O from $-20°$ C to $140°$ C and from 0 kPa to 140 kPa. The pressure at which ice and water coexist in equilibrium falls very rapidly with increasing temperature, by 10 000 kPa for every $1°$ C. Within the range of pressures shown in the figure, therefore, the variation in temperature is about $1/100°$ C, too little to show on the diagram. Thus the ice-water transition is effectively isothermal in the kind of simple experiment illustrated in Figure 4.3. The univariant state of the system in Figure 4.3 (b) can therefore be regarded as one in which a change of temperature alters the system from a two-phase condition to a one-phase condition, while a change of pressure has no effect. This may alternatively be expressed by saying that the system has one degree of freedom because the pressure can be changed without altering the state of the system. This univariant change in pressure only is indicated by the arrows on the ice-water univariant curve.

mined from outside the rock — temperature by proximity to an intrusion or the regional heat flow, and pressure by the depth of burial. So for natural recrystallising metamorphic rocks the number of degrees of freedom is likely to be equal to or greater than two. Equation (1) can be modified for metamorphic rocks to be:

$$P \leq C \tag{2}$$

P is now simply the number of minerals in the assemblage list, C the number of chemical components in the rock composition. C can be determined by finding out how many oxide ratios are needed to describe the compositions of all the minerals. This relationship was first demonstrated by Goldschmidt in the Oslo hornfelses, and is known as **Goldschmidt's mineralogical Phase Rule**. Most modern petrogenetic schemes for metamorphic rocks are based on the application of this rule (or, to be more accurate, a modification of it to allow for the presence of a vapour phase during metamorphism in many rocks). It involves the much-discussed concept of metamorphic facies which will be briefly considered in Chapter 13.

COMPOSITION–ASSEMBLAGE DIAGRAMS

One very useful application of the Phase Rule to metamorphic rocks is that it permits the preparation of diagrams which represent the mineral assemblages of a range of different rock compositions.

The problem in preparing such diagrams is that the number of major chemical components in most metamorphic rocks is five or six, while a two-dimensional diagram can only represent phase relationships in a system with at most three components. Attempts have been made to use perspective drawings of three-dimensional composition–assemblage diagrams, but these are hard to visualise and even harder to memorise. The value of diagrams lies in their being an aid to understanding and remembering mineral assemblages, so in most cases it is probably not worth the struggle to work with more than three components.

Since Goldschmidt first used this method, there have been many ingenious attempts to devise composition–assemblage diagrams to describe most metamorphic rock compositions. Goldschmidt's original ACF triangular diagrams are among the most successful; they are described and used in this book. But as knowledge of mineral assemblages has increased, and as the metamorphism of a wider range of rock compositions is investigated, such general diagrams become less adequate.

In this book therefore, composition–assemblage diagrams applicable to *all* metamorphic rocks will not be employed. Since two main rock compositions will be discussed, two types of diagram particularly appropriate for these compositions will be used. On occasion, other diagrams will be used for special cases for which they are appropriate. They will be explained when they are used. The two types which will be specially described in this chapter are the **AFM diagram** for pelitic sediments and the **ACF diagram** for basic igneous rocks.

The reader who is interested in further discussion of these and other types of composition–assemblage diagrams is referred to Winkler (1976, Ch. 5).

The AFM diagram

AFM diagrams are used to represent the rock compositions and mineral assemblages of pelitic rocks. They were invented by Thompson (1957). Six major components of the rock composition are considered: SiO_2, Al_2O_3, Fe_2O_3, FeO, MgO and K_2O. The minor components Fe_2O_3, TiO_2 and P_2O_5 are also considered. Pelitic rocks usually contain free quartz (Ch. 5), indicating that there is an excess of SiO_2 over that needed to form the other silicate minerals in the assemblage list. Therefore, provided that the caption to the diagram states that all the mineral assemblages represented include quartz, SiO_2 need not be considered further. As well as quartz, for an AFM plot to be used, all the mineral assemblages must include either muscovite or K-feldspar. These two minerals are rarely found together in pelitic rocks, although they are in metamorphosed acid igneous rocks.

If the rocks are muscovite bearing, a Thompson AFM diagram is made. The molecular proportions of $[Al_2O_3]$, $[FeO]$, $[MgO]$ and $[K_2O]$ are calculated from the weight percentage list. The relative amounts of these four components fix positions inside a tetrahedral volume, which represents the rock compositions. In Figure 4.5, X, Y and Z represent three different rock compositions. X is relatively rich in Al_2O_3, Y is poor in Al_2O_3 and rich in K_2O and Z contains no K_2O. X and Y are represented by points within the volume of the tetrahedron, Z by a point on its lower face. The Thompson triangular diagram projects compositions from the interior of the tetrahedron onto the lower $[Al_2O_3]$, $[FeO]$, $[MgO]$ face (or its extension outside the tetrahedron). The projection is made along lines from the point representing the composition of pure muscovite ($K_2Al_4Si_6Al_2O_{20}$. $(OH)_4$, i.e. K_2O .$3Al_2O_3$.$6SiO_2$.$2H_2O$) on the $[Al_2O_3]$ - $[K_2O]$ edge of the tetrahedron, through the rock composition onto the $[Al_2O_3]$ - $[FeO]$ - $[MgO]$ face. The names of the chemical components are abbreviated for convenience to A for $[Al_2O_3]$, F for $[FeO]$ and M for $[MgO]$. Some corrections are made to the major components in the analysis before plotting to allow for the presence of the components TiO_2, Fe_2O_3, and Na_2O. TiO_2 is assumed to be present as ilmenite ($FeTiO_3$) in the 'opaques' of the assemblage list, and Fe_2O_3 in magnetite (Fe_3O_4).

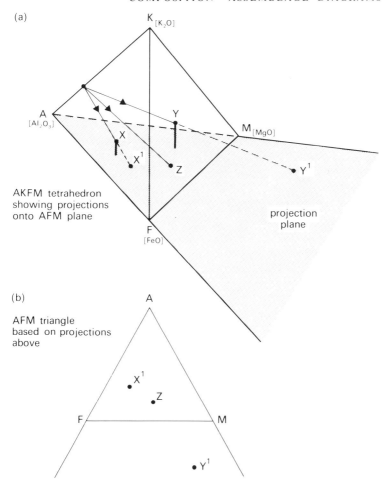

Figure 4.5 Projection of rock and mineral compositions onto the base of an AKFM tetrahedron to form a Thompson AFM triangular diagram.

A worked example with instructions for plotting follows. Exercises in plotting rock compositions on composition–assemblage diagrams are also given at the end of each chapter of Part II. The calculations are intended to be performed on a pocket calculator, and the student is advised to work through the arithmetic of this example for himself. The cheapest type of 'four function' calculator is adequate for all the calculations in this book, although a 'memory' facility will help in some cases.

Table 4.1 Calculation of AFM values for an analysis of a pelitic rock, plotted on Figure 4.6a.

	A	B	C
SiO_2	53·0	60·07	0·8823
TiO_2	0·9	79·89	0·0113
Al_2O_3	22·1	101·82	0·2170
Fe_2O_3	2·1	159·68	0·0132
FeO	1·8	71·84	0·0251
MnO	0·01	70·93	0·0001
MgO	1·7	40·31	0·0422
CaO	0·5	56·07	0·0089
Na_2O	0·6	61·97	0·0097
K_2O	6·3	94·20	0·0669
P_2O_5	0·3	141·92	0·0021
H_2O (total)	5·1	18·01	0·2832
C	5·3	12·01	0·4413
Total	99·7		

Calculation of F

$F = 0·0251 - 0·0113 - 0·0132 = 0·0006$. $F\% = 0·0006/0·0591 \times 100 = 1·0\%$

Calculation of A

$A = 0·2170 - 3 \times 0·0669 = 0·0163$. $A\% = 0·0163/0·0591 \times 100 = 27·6\%$

$A + F + M = 0·0163 + 0·0006 + 0·0422$
$= 0·0591$. $M\% = 0·0422/0·0591 \times 100 = 71·4\%$

Column A: weight percentages of oxides in Carboniferous phyllite from Lötschental, Switzerland. Analysed by E. Regli for C. Taylor.
Column B: molecular weights, see Appendix.
Column C: molecular proportions, for derivation see the text.

The numbers for the worked examples are given in Table 4.1. Proceed as follows:

(1) Divide the weight percentages of the oxides in the analysis (Column A) by their molecular weights (Column B). A list of molecular weights of oxides and elements given in rock analyses is to be found in the Appendix. The results of the divisions are the molecular proportions listed in Column C.

(2) $[FeO]$ must be corrected to allow for the ferrous iron in the ilmenite and magnetite of the opaque minerals of the assemblage, which are not included in the AFM plot. Ilmenite has the formula $FeO.TiO_2$, and magnetite the formula $FeO.Fe_2O_3$. Therefore $F = [FeO] - [TiO_2] - [Fe_2O_3]$.

(3)   The projection from pure muscovite composition onto the base of the AFMK tetrahedron is performed as follows. The chemical formula of pure muscovite may be written in the form $K_2O.3Al_2O_3.6SiO_2.2H_2O$. Therefore three times the molecular proportion of K_2O is subtracted from the molecular proportion of Al_2O_3, i.e. $A = [Al_2O_3] - 3[K_2O]$.

(4) $M = [MgO]$.

(5)   The totals A, F and M are added. Each is expressed as a proportion of the total. It is conventional to multiply the proportions by 100 to express the result in percentages.

(6)   Plot the result on triangular graph paper. The conventional positions for A, F and M, and the position of the rock composition in the worked example, are shown in Figure 4.6a.

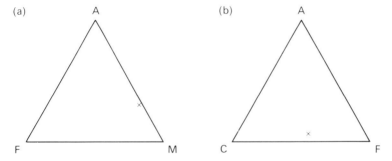

Figure 4.6   Rock compositions plotted on triangular diagrams (a) Thompson AFM plot of pelitic rock from table 4.1. (b) ACF plot of basic igneous rock from table 4.2.

Sometimes $[MnO]$ is added to the $[FeO]$ total in the calculation of F. This is because Mn^{+2} substitutes for Fe^{+2} in many silicate minerals. This may occasionally lead to trouble if Mn-rich garnets are present in the mineral assemblage (Ch. 15) and has therefore not been recommended here. The proportion of $[MnO]$ in most pelitic rocks is very small, so the differences between including it in F and leaving it out are not great.

The ACF diagram

ACF diagrams are used to represent the rock compositions of rocks of basic igneous composition and are also useful for impure limestones. Five of the

Table 4.2 Calculation of ACF values for an analysed basic igneous rock, plotted on Figure 4.6b.

	A	B	C
SiO_2	48·62	60·07	0·8094
TiO_2	1·84	79·89	0·0230
Al_2O_3	9·71	101·82	0·0954
Fe_2O_3	5·40	159·68	0·0338
FeO	4·45	71·84	0·0619
MnO	0·16	70·93	0·0023
MgO	7·69	40·31	0·1908
CaO	9·89	56·07	0·1764
Na_2O	3·94	61·97	0·0636
K_2O	0·31	94·20	0·0033
H_2O (total)	6·80	18·01	0·3776
P_2O_5	0·18	141·92	0·0013
CO_2	0·86	44·00	0·0195
Total	99·85		

$A = Al_2O_3 - Na_2O - K_2O = 0·0954 - 0·0636 - 0·0033 = 0·0285$
$C = CaO - 10/3 \times P_2O_5 - CO_2 = 0·1721 - 0·0043 - 0·0195 = 0·1526$
$F = MgO + FeO + Fe_2O_3 + TiO_2 = 0·1908 + 0·0619 + 0·0338 + 0·0230 = 0·1959$
$A + C + F = 0·3770$
$A = 7·6\%$ $C = 40·4\%$ $F = 52·0\%$

Column A: weight percentages of oxides in a metamorphosed basalt from Mutki, Turkey, analysed by R. Hall.
Column B: molecular weights from Appendix.
Column C: molecular proportions calculated as explained in text.

major components are considered: SiO_2, Al_2O_3, FeO, MgO and CaO. In this case, unlike the AFM diagram, FeO and MgO are assumed to substitute for one another perfectly in mineral structures. The assumption that all mineral assemblages are saturated with SiO_2 and therefore contain quartz in the mineral assemblages is shared with the AFM diagram. Three components are therefore left for plotting: $[Al_2O_3]$, $[MgO] + [FeO]$, and $[CaO]$. Once again corrections are applied for minor elements. $[FeO]$ is modified as in the AFM diagram for $[TiO_2]$ and $[Fe_2O_3]$. $[CaO]$ is corrected for the proportion which combines with $[P_2O_5]$ in apatite. The formula of apatite may be written $10CaO.3P_2O_5$ so $10/3 \times [P_2O_5]$ is subtracted from the $[CaO]$ total. $[Na_2O]$ and $[K_2O]$ are assumed to be present in plagioclase feldspar ($[K_2O]$ as the small amount of solid solution possible in plagioclase). Since the formula of albite may be written $Na_2O.Al_2O_3.6SiO_2$, and of orthoclase $K_2O.Al_2O_3.6SiO_2$, $[Na_2O]$ and $[K_2O]$ are subtracted from the $[Al_2O_3]$ total, assuming that plagioclase is always present in the assemblage list. This leaves three components, A = $[Al_2O_3]$, C = $[CaO]$ and F = $[MgO] + [FeO]$. Notice that 'A' and 'F' of the ACF diagram are different from 'A' and 'F' of the AFM diagram.

Once again a worked example is given, the calculation being shown in full in Table 4.2. Proceed as follows:

(1) Divide weight percentages (Column A) by molecular weights (Column B) to obtain molecular proportions (Column C) as in the AFM calculation.

(2) Correct $[Al_2O_3]$ by subtracting the aluminium present in feldspars. A = $[Al_2O_3] - [Na_2O] - [K_2O]$.

(3) Correct CaO by subtracting the calcium present in apatite $(10CaO.3P_2O_5)$ and in calcite $(CaO.CO_2)$.

$$C = [CaO] - 10/3 \times [P_2O_5] - [CO_2].$$

(4) Correct $[FeO]$ by subtracting the ferrous iron in ilmenite and magnetite:

$$[FeO] = [FeO] - [TiO_2] - [Fe_2O_3].$$
$$F = [MgO] + [FeO]$$

(5) Find the sum of A + C + F.
(6) Express A, C and F as a percentage of (A+C+F).

The conventional positions of A, C and F, and the position of the rock plotted in Table 4.2, are shown in Figure 4.2b.

Mineral compositions on AFM and ACF diagrams

Because minerals usually have constant proportions of oxides, they usually plot as points on composition–assemblage diagrams. This has been illustrated already by the composition of muscovite in the AKFM tetrahedron. Exceptions to this rule are minerals where one major element may substitute for another in the crystal structure (e.g. Fe^{+2} for Mg in orthopyroxenes). These will plot as lines if there is only one ionic substitution, as in orthopyroxenes, or as areas on the diagram if there is more than one substitution possible, as in hornblende on an ACF diagram. The length of the line, or the size of the area, illustrate the range of possible substitution.

Within the wide range of compositions which can be indicated on the

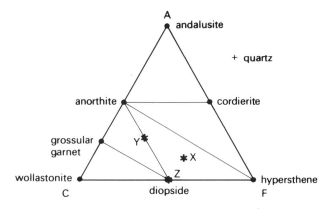

Figure 4.7 Example of an ACF diagram with tie-lines.

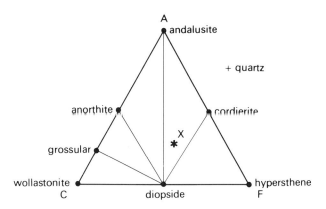

Figure 4.8 An ACF diagram with nonsense tie-lines (see text).

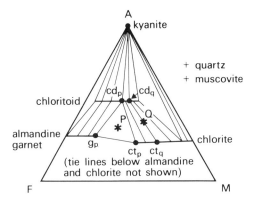

Figure 4.9 Example of a Thompson AFM diagram with tie-lines.

whole triangular diagram, several different mineral assemblages may occur. Remember that, because the diagram can only represent three components, the Phase Rule says that only assemblages with one, two or three minerals can be shown. Other minerals may be present in addition for all compositions shown (e.g. muscovite and quartz in the AFM triangle).

If none of the minerals show substitutions of the components plotted on the diagram, a simple diagram such as that of Figure 4.7 is obtained. The lines joining mineral compositions are called **tie-lines**. A rock of composition X will be represented by the four-mineral assemblage anorthite + hypersthene + diopside + quartz (the quartz is present because there is an excess of SiO_2 in the rock, as explained earlier). A rock of composition Y will have the three-mineral assemblage anorthite + diopside + quartz. A rock of composition Z will have the two mineral assemblage diopside + quartz.

The pattern of tie-lines in Figure 4.7 does not follow from the mineral compositions alone. Figure 4.8 shows a different distribution of tie-lines. This is a nonsense diagram: the assemblages it indicates e.g. andalusite + diopside + cordierite + quartz for composition X, do not occur in real rocks. Many triangular diagrams will be used in this book and the basis of the tie-line patterns from mineral assemblages will often be discussed.

If ionic substitutions are possible, the diagrams become more complicated. Because the substitution $Fe^{+2} \rightleftharpoons Mg$ is so common, minerals are often represented by horizontal lines rather than points on AFM diagrams. An example is shown in Figure 4.9. The chloritoid, garnet and chlorite in Figure 4.9 have limited ranges of substitution of Fe^{+2} for Mg. A rock of composition P has the assemblage chloritoid + chlorite + garnet + muscovite + quartz. The compositions of the chloritoid, chlorite and garnet in this rock are cd_p, ct_p and g_p. A rock of composition Q has the mineral assemblage chloritoid + chlorite + muscovite + quartz. The chloritoid in this rock has the composition cd_q and the chlorite ct_q. For both compositions, P and Q, the compositions of minerals which coexist are joined by tie-lines. Composition Q lies in a field over which only two mineral compositions are represented on the diagram for each rock composition. The tie-line cd_q–ct_q is one of an infinite number of possible tie-lines, depending on the ratio of Fe^{+2} to Mg in the rock. The lines shown on the AFM triangle are a few representative members of the family of tie-lines crossing the two-mineral field.

It is possible to determine for a particular rock the composition of the rock and of all the Fe- and Mg-bearing minerals, and put an AFM triangle such as that of Figure 4.9 onto a quantitative basis (Ch. 15). But the diagram can also be useful for illustrating the possible mineral assemblages without exact knowledge of mineral composition ranges, and such semi-quantitative diagrams will often be used in this book. The bounding tie-lines (e.g. cd_p – g_p – ct_p) are of great interest in petrogenetic studies. All the diagrams show the equilibrium metamorphic mineral assemblages for a particular range of metamorphic conditions only. The patterns of tie-lines change with metamorphic grade and one use of triangular diagrams is to represent the change of mineral assemblages for a range of rock compositions in progressive metamorphic sequences. They can summarise the differences in metamorphic assemblages between rocks of different areas. The AFM triangles of Figure 5.5 illustrate a change in mineral assemblages with metamorphic grade in one contact metamorphic aureole, and comparison of the ACF triangles in Figure 8.13 with those of Figure 11.6 summarises the differences in metamorphic assemblages between Sulitjelma, Norway, and the Troodos mountains, Cyprus.

5

Contact metamorphic rocks

Rocks of pelitic composition show most striking changes in texture and mineral assemblage in contact aureoles. The example to be described here has been the type contact aureole presented to British geology students for more than sixty years. It was described by Rastall (1910).

THE SKIDDAW CONTACT AUREOLE

In high moorland country in the northern part of the English Lake District, in the county of Cumbria, several small areas of outcrop of biotite granite or granodiorite are seen in contact with Ordovician Skiddaw Slates. Radiometric age determinations indicate that the intrusions are of Devonian age. They are surrounded by a wide contact aureole, and study of the contacts suggests that they join at a small depth below the land surface, representing the domed roof of a medium-sized granitic batholith (Fig. 5.1).

Outside the contact aureole, the slates have a penetrative cleavage associated with tight folding. Cleavage and folds are overprinted by the contact aureole, indicating that they belong to a pre-intrusive episode of regional metamorphism and rock deformation. Therefore the Skiddaw Aureole shows contact metamorphism superimposed on low-grade regional metamorphism.

Microscopic study of the Skiddaw Slates outside the contact aureole reveals mineral assemblages of low-grade regional metamorphism. The phyllosilicate minerals are muscovite and chlorite, with a strong preferred orientation parallel to the **slaty cleavage** direction. Grains of quartz showing their original clastic shape, small prisms of chloritoid and opaque minerals, iron sulphide, iron oxide and graphite are present. An unresolved problem is that the Skiddaw Slates are relatively rich in aluminium, but some do not contain chloritoid, and in these the aluminium-rich mineral has not been identified. It has been suggested that it might be an aluminium-rich phyllosilicate end-member in the 'muscovite', which would therefore be sericite rather than muscovite (Ch. 10) but this has not been checked by X-ray mineral determination.

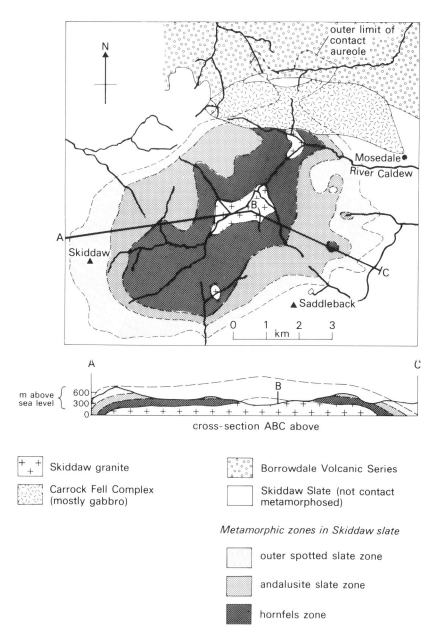

Figure 5.1 Geological sketch map of the Skiddaw Granite and its contact aureole, from Eastwood *et al*. (1968).

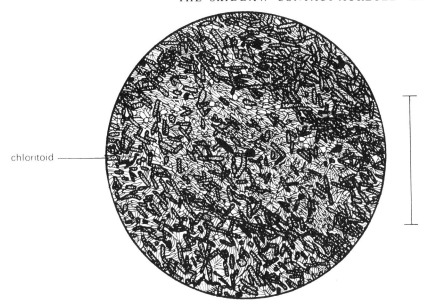

chloritoid

Figure 5.2 Spot in spotted slate, outer aureole of Skiddaw. Scale bar 1 mm.

The progressive sequence in the pelitic rocks of the Skiddaw Aureole has been divided into three zones, giving the sequence in order of increasing metamorphism: unaltered Skiddaw Slates → outer spotted slate zone → intermediate porphyroblastic slate zone → inner hornfels zone. These zones are distinguished on a textural basis rather than on variations in the mineral assemblages. The outer spotted slate zone is characterised by a coarsening of the grain size of the slate and a tendency for the cleavage to be less perfect. Spots $0.15–2.0$ mm across appear on the cleavage surfaces, but they cannot be identified mineralogically in the field. In the intermediate porphyroblastic slate zone, the porphyroblasts can be identified as andalusite (a distinctive variety known as **chiastolite**) and as cordierite. The grain size is coarser, to the extent that some of the rocks might be described as fine-grained phyllites, but the cleavage is still present. In the inner hornfels zone the rocks have become massive, although the cleavage can often still be recognised.

Figure 5.2 is a sketch of a thin section through a spot in the outer spotted slate zone. The section is cut at right angles to the foliation, as are all thin sections of foliated rocks illustrated in this book. It can be seen from the sketch that the spot contains no different minerals from the surrounding matrix. Both consist of the assemblage muscovite + chlorite + chloritoid + opaque minerals. The lighter-coloured spot contains more muscovite and less chlorite and chloritoid than the matrix, but the grain size throughout the rock is uniform. Spots of this kind, not showing any different minerals from the surrounding matrix, are quite a common feature in the outer part of contact aureoles in pelitic sediments. Their mode of origin is uncertain. Some authors, following Rastall's classic view, suggest that they are due to selective concentration of certain chemical components which might have preceded the formation of porphyroblasts of cordierite or andalusite; but it is also possible that they represent **retrograde metamorphism** of porphyroblast minerals. The higher-grade parts of the Skiddaw Aureole certainly show partial replacement of andalusite crystals by muscovite, and occasionally total replacement. The chloritoid + chlorite + muscovite spot in Figure 5.2 could have formed by retrogression of cordierite, by hydration and introduction of K^+ ions.

Many of the slates of the outer spotted slate zone contain biotite, which occurs in rocks of suitable composition right out to the edge of the zone. In the inner parts of the outer spotted slate zone, porphyroblasts of andalusite and cordierite have been recognised in thin section, although they are not identifiable in the field.

Figure 5.3 shows a cordierite-andalusite slate from the intermediate porphyroblastic slate zone. It consists of the assemblage biotite + muscovite + cordierite + andalusite + quartz + opaque minerals. The cordierite forms oval crystals with irregular outlines and numerous inclusions

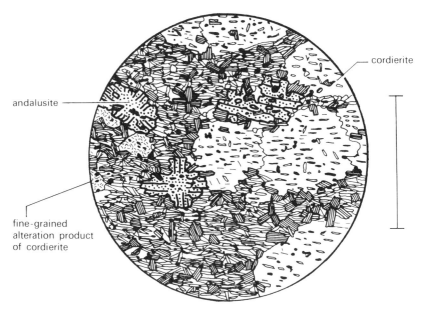

Figure 5.3   Cordierite-andalusite slate from intermediate zone, Skiddaw Aureole. Scale bar 1 mm.

of biotite, muscovite and opaques. This is an example of **poikiloblastic texture**, which is the name given when crystals of metamorphic minerals contain inclusions of other minerals. In this case, the inclusions are so numerous that a special name, **sieve texture**, is sometimes used. Notice that the biotite and muscovite flakes, and also flaky opaque grains (which are probably graphite) show a strong parallel preferred orientation even within the cordierite crystals. This is the **relict** slaty cleavage of the Skiddaw Slates, and this observation that it is overgrown by cordierite is part of the evidence indicating that the contact metamorphism of the Skiddaw aureole is later than the formation of the cleavage in the slates. The andalusite and cordierite crystals are **porphyroblasts**, coarser grained crystals in a finer grained groundmass, and the drawing shows one prismatic section and two near basal sections through prismatic andalusite crystals. The basal sections at a first glance appear to be twinned crystals, with re-entrant grain boundaries at the corners and a different cleavage direction visible in different parts of the crystal. They are not twinned, however, as examination between crossed polars reveals, and their textural features are due to an early stage of development of the **chiastolite** habit in the andalusite, which is more fully shown by the next thin section, from the inner hornfels zone.

A hornfels from the inner hornfels zone is shown in Figure 5.4. The mineral assemblage is the same as that in the hornfels of the intermediate

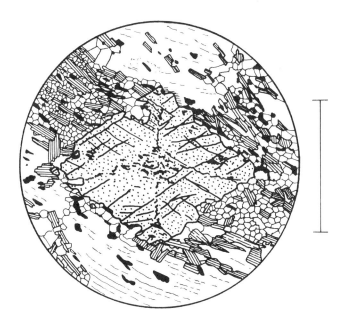

Figure 5.4 Andalusite-cordierite schist from inner zone, Skiddaw Aureole. Scale bar 1 mm.

zone, biotite + muscovite + cordierite + andalusite + quartz + opaque minerals. The original slaty cleavage is still visible as preferred orientation of biotite, muscovite and graphite flakes. The cordierite crystals are clear of muscovite and biotite inclusions, but still have graphite flakes and trails of equant opaque minerals showing that they have overgrown the slaty cleavage. The andalusite porphyroblast is shown in cross section at right angles to the length of the prism and has a distinctive chiastolite cross of opaque inclusions. The cross is a product of the growth of the andalusite porphyroblast and, by comparing the andalusite crystals in Figures 5.3 and 5.4 it is possible to see how it has arisen. The porphyroblasts have grown by replacing muscovite, biotite and quartz selectively along grain boundaries, as can be seen from the irregular shape of the boundary in both figures. However, in the growing crystals, boundaries parallel to the prism direction {110} are more stable than others, because they have a lower surface energy. Such boundaries tend to predominate and crystals of other minerals cutting them have been replaced more readily than those on the corners of the growing prism. As a result, the directions of growth of the edges of the prismatic crystal are marked by trails of opaque inclusions, which constitute the chiastolite cross. The reason for the preferred development of one direction of cleavage in the sectors between the trails of inclusions is not known.

This hornfels is coarser-grained than the rocks of the outer aureole and the fabric in the quartz crystals is easily seen. The quartz grains are equidimensional and in cross section approximate to a mosaic of approximately equilateral 5-, 6- or 7-sided polygons. The distinctive feature of the texture of the quartz is that the triple junctions seen in thin section tend to intersect at 120°. This is the **granoblastic texture** typical of hornfelses. In hornfelses without prismatic minerals such as andalusite and platy minerals such as muscovite and biotite, all the minerals in the rock have this texture. It is similar to the textures found in metals which have been annealed by heating without any deformation, and implies that grains of similar size and with grain boundaries of equal surface energy have attained a state of textural equilibrium. Textural equilibrium usually also indicates equilibrium in the Phase Rule sense of Chapter 4 and therefore mineral grains in contact in a texture of this kind are members of an equilibrium mineral assemblage.

The sequence of mineral reactions in the Skiddaw Aureole is not easy to determine, because the aureole has been mapped on textural variations in the rocks. Most of the significant reactions seem to occur in the outer spotted slate zone. It is possible to estimate the sequence in which the new minerals appear with increasing metamorphic grade, but it is virtually impossible to determine the sequence in which minerals such as chlorite and chloritoid disappear, because of the evidence for a degree of retrograde metamorphism mentioned earlier.

The new minerals appearing are first biotite and, shortly after at nearly

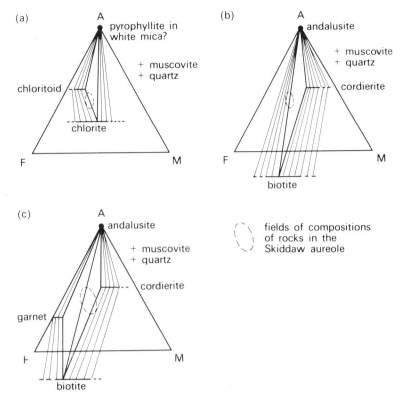

Figure 5.5 AFM diagrams illustrating the progressive metamorphism of Skiddaw Slates in the Skiddaw Aureole. (a) Mineral assemblages of Skiddaw Slates and outer aureole. (b) Mineral assemblages of intermediate zone. (c) Mineral assemblages of inner hornfels zone.

the same metamorphic grade, cordierite and andalusite. The incoming of these minerals is accompanied by the disappearance of chlorite and chloritoid in most rocks (compare Figures 5.2 and 5.3). Thus the sequence of assemblages in typical rocks of the Skiddaw Aureole is as follows:

chlorite + chloritoid + sericite + quartz	Unaltered slates
biotite + chlorite + sericite + quartz	Low-grade assemblages
chloritoid + biotite + sericite + quartz	
cordierite + andalusite + muscovite +	High-grade
biotite + quartz	assemblages

In addition, Rastall (1910) recorded garnet in a few rocks, presumably of unusual compositions, in the higher-grade parts of the aureole. The garnet has been shown to have a very manganese-rich composition (Tilley 1926).

Without more knowledge of the compositions of minerals, especially white mica and chlorites, than is available at present, it is not possible to write chemical equations for the formation of biotite, cordierite and andalusite. But because all assemblages contain quartz and muscovite, they can appropriately be shown on AFM diagrams (Fig. 5.5). The diagrams are schematic because the amount of $Fe^{+2} \rightleftharpoons Mg$ substitution in cordierite, chlorite, etc., is not known, but they give some idea of the assemblages observed.

THE COMRIE AUREOLE

To the north-west of the Highland Boundary Fault, in Perthshire, Scotland, an intrusion of diorite and pegmatite cuts Dalradian schists. There is a distinct contact aureole surrounding the intrusion (Fig. 5.6).

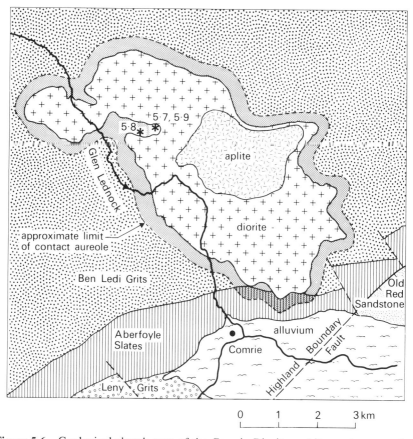

Figure 5.6   Geological sketch map of the Comrie Diorite and its contact aureole. The limits of the aureole have not been mapped, but have been put at the approximate outer distance given by Tilley (1924a). Locations of samples in Figures 5.7, 5.8 and 5.9 indicated.

As in the Skiddaw Aureole, the contact metamorphism of the Comrie Diorite is overprinted upon low-grade regional metamorphism associated with the development of a penetrative cleavage in the pelitic rocks. The regional metamorphism belongs to the chlorite zone of the progressive metamorphic sequence described by Barrow (1912). The mineral assemblages of the contact aureole of the Comrie Diorite were described in detail by Tilley (1924a).

The diorite cuts two formations of the Dalradian Supergroup, the Aberfoyle Slates and the Ben Ledi Grits. (Fig. 5.6). The Aberfoyle Slates are aluminium-rich pelitic rocks, and display a similar contact metamorphic sequence to that of the Skiddaw Aureole, the distance from the first appearance of spots on the slates to the contact being 410 m. This sequence will not be discussed further here.

The Ben Ledi Grits are metamorphosed turbidites and have a greater variation in chemical composition than the Aberfoyle Slates or the Skiddaw Slates in the Skiddaw Aureole. The feature of interest is that while the metamorphic assemblages of the Ben Ledi Grits are usually rich in SiO_2, some of the hornfelses close to the contact with the diorite are very poor in SiO_2 and quartz does not occur in their metamorphic assemblages. Silica-undersaturated minerals such as spinel and corundum are found instead. Tilley attributed this to local metasomatism at the contact, which depleted the country rocks in SiO_2. Another possibility is that local partial melting occurred close to the contact. Partial melting would tend to produce an SiO_2-rich liquid of granitic composition and the unmelted residue would be correspondingly depleted in SiO_2. Also, perhaps because the diorite magma of the Comrie intrusion was hotter than the granite magma of the Skiddaw intrusion, in the inner part of the Comrie Aureole muscovite is absent from the mineral assemblages, its place being taken by orthoclase feldspar.

Figure 5.7 is a drawing of a thin section of a corundum and spinel-bearing hornfels from the inner part of the aureole. The corundum is colourless with high relief. Between crossed polars it shows interference colours of the first order (up to yellow) and straight extinction parallel to the direction of preferred orientation of the minute opaque inclusions. The spinel is isotropic and also distinctive in plane-polarised light because of its deep emerald green colour. Its relief is also high, but a Becke line test shows that it has a lower R.I. than the corundum. This green spinel is of an intermediate composition between spinel in the strict sense ($MgAl_2O_4$) and hercynite ($Fe^{+2}Al_2O_4$). Spinel of this intermediate composition and green colour is quite often found in high-grade contact metamorphic pelitic rocks, and is called **pleonaste**. The rock also contains an opaque mineral, probably magnetite (Fe_3O_4) which is also a member of the spinel family of minerals. There is a difference in crystal structure between pleonaste and magnetite, pleonaste being a normal spinel and magnetite an inverse spinel (Deer *et al.*

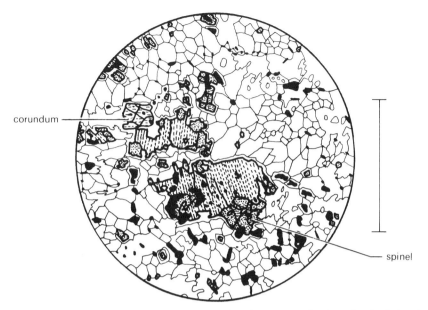

Figure 5.7 Very SiO₂ under-saturated hornfels, inner aureole, Comrie. Scale bar 1 mm.

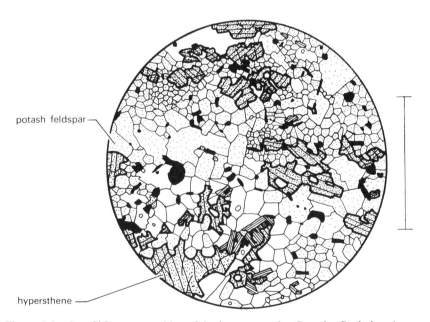

Figure 5.8 Just SiO₂ saturated hornfels, inner aureole, Comrie. Scale bar 1 mm.

1966, p. 425). The coexistence of the two together in the mineral assemblage indicates that the amount of solid solution of inverse spinel into normal spinel and vice versa is very limited.

The mineral assemblage of this hornfels is: cordierite + orthoclase + corundum + spinel + biotite + magnetite. The cordierite in the thin section is almost impossible to distinguish from plagioclase feldspar of the composition range An_{20}–An_{30}. Refractive indices, birefringence and optic signs are sufficiently close to prevent distinction by the petrological microscope. The oblique sector twins and sieve texture which are a distinctive feature of the cordierite of the Skiddaw Aureole are not so well developed at Comrie. The cordierite shows polysynthetic twinning. Careful study, however, shows occasional twin lamellae intersecting at 60°. This does not occur in plagioclase. The alteration of the cordierite along cracks to a distinctive yellow alteration product is also diagnostic. The best confirmatory test for cordierite is to find yellow pleochroic haloes round inclusions of radioactive minerals. The orthoclase is easy to recognise because its R.I. is lower than that of the mounting medium of the thin section.

The grain size of this rock is coarser than that of the hornfels from the inner part of the Skiddaw Aureole (Fig. 5.4). The cordierite and orthoclase have the same granoblastic texture as that of the quartz in the Skiddaw hornfels. The clusters of corundum and spinel crystals often have a narrow rim of cordierite against orthoclase.

Figure 5.8 shows a hornfels from the inner part of the Comrie Aureole which contains neither quartz nor silica-under-saturated minerals. The commonest ferromagnesian mineral is hypersthene, occurring as skeletal porphyroblasts in a finer-grained groundmass of orthoclase and cordierite. The mineral assemblage is: hypersthene + cordierite + orthoclase + biotite + opaques. This type of rock is very common in the inner parts of contact aureoles and is called **pyroxene hornfels**.

Figure 5.9 shows a hornfels from the inner part of the aureole containing abundant quartz. The ferromagnesian mineral here is again different. It is the magnesium-iron amphibole cummingtonite. The assemblage in this rock is: quartz + plagioclase (An_{38}) + biotite + cummingtonite + opaques. The cummingtonite occurs as radiating clusters of prisms. It has polysynthetic twinning, with an oblique extinction of 18° on the twin composition planes. The clusters of cummingtonite crystals may be **pseudomorphs** replacing original hypersthene, but it seems reasonable in view of the general equilibrium in the Comrie hornfelses, to regard cummingtonite as an equilibrium mineral. It is therefore included in the assemblage list above.

The composition of the hornfelses of the inner part of the Comrie Aureole may be represented on a special triangular diagram designed to illustrate the degree of over- or under-saturation by SiO_2 and the relative proportions of Al_2O_3 and of $FeO+MgO$ taken together (Figure 5.10). This

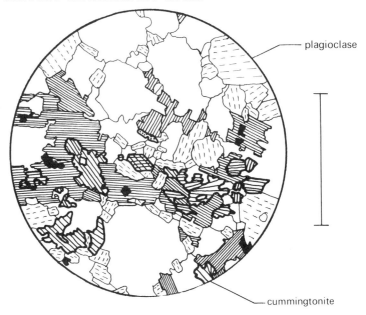

plagioclase

cummingtonite

Figure 5.9 SiO₂ over-saturated hornfels with quartz and cummingtonite, inner aureole, Comrie. Scale bar 1 mm.

type of diagram is suggested by the discussion of the hornfelses in Tilley (1924a) although he did not actually plot one. The approximate range of composition of the Skiddaw Slates, which is probably similar to that of the Aberfoyle Slates, is indicated. The Ben Ledi Grit compositions extend over a large part of the diagram. The **tie-lines** from andalusite to cordierite and from cordierite to cummingtonite are emphasised. In the sub-triangles below and left of these lines the equilibrium assemblages contain quartz while in the rest of the triangle the assemblages are under-saturated, except in the small triangle cummingtonite–cordierite–hypersthene, where the rocks contain neither quartz nor undersaturated minerals. Estimated compositions of the rocks shown in Figures 5.7, 5.8 and 5.9 are shown.

In the Skiddaw and Comrie Aureoles, the hydrous minerals such as chlorite, chloritoid and muscovite of the low-grade rocks outside the aureole are replaced in the hornfelses of the inner aureole by virtually anhydrous minerals such as cordierite, hypersthene and andalusite. The process of contact metamorphism may therefore be described as **progressive dehydration**. Most of the specific dehydration reactions which occur in pelites are too complex to discuss here, and in Skiddaw are not known because of lack of detailed mineralogical study. One reaction which occurs in the Comrie Aureole is simple and will be discussed here and, in more detail, in Chapter 13. In the hornfelses of the Comrie Aureole, muscovite has broken down with increasing metamorphic grade and its place has been taken by orthoclase.

Two different chemical reactions may account for this change, depending whether the rock is over- or under-saturated with SiO_2. If it is over-saturated, the reaction is muscovite + quartz $\rightleftharpoons$ orthoclase + andalusite + H_2O. If it is under-saturated the reaction is muscovite $\rightleftharpoons$ orthoclase + corundum + H_2O. In both cases the H_2O is driven off as a vapour, presumably escaping along grain boundaries and fractures in the rocks.

The reactions are directly reflected in changes in the mineral assemblage lists which may contain other minerals as well as the ones in the equations above. At Comrie, with both over- and under-saturated rocks containing orthoclase but not muscovite, both reactions must have occurred. At Skiddaw the first reaction has not occurred. A glance at Figure 13.4 shows that the minimum temperature difference between the hornfelses of the two aureoles can be estimated. If the pressure in each case was about 0.1 GPa (representing a depth of burial of 3–4 km) the temperature in the inner part of the Comrie Aureole must have exceeded 700 °C; that in the inner part of the Skiddaw Aureole cannot have exceeded 600 °C.

Both the contact aureoles described so far are in country rocks of pelitic composition. Contact metamorphism of basic igneous rocks will not be described in detail, because the mineral assemblages in the lower-grade

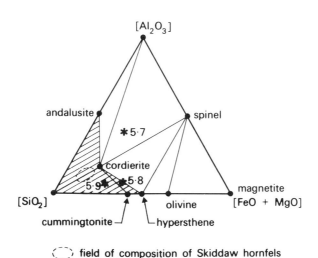

Figure 5.10 $[Al_2O_3] - [SiO_2] - [FeO + MgO]$ triangular diagram showing mineral assemblages of over- and under-saturated hornfelses, Comrie. Shaded area, over-saturated assemblages. Approximate compositions of rocks in Figures 5.7, 5.8 and 5.9 indicated, and also field of composition of Skiddaw aureole rocks.

parts of the aureoles are rather like those of progressive regional metamorphic sequences, which will be described later. In the inner, high-grade parts of contact aureoles, basic igneous rocks commonly show mineral assemblages such as clinopyroxene + plagioclase (An_{50}) + quartz + opaques. This is similar to the list of minerals which might be found in a basalt. The texture of a hornfels of basic igneous composition is grano-blastic and, although olivine-bearing basalts are quite common, olivine is not usually found in contact metamorphosed basalts. These rocks, like orthopyroxene-rich metamorphosed pelitic sediments, are called **pyroxene hornfelses**.

THE BEINN AN DUBHAICH AUREOLE, ISLE OF SKYE

Contact metamorphism of carbonate-rich sediments introduces an important new factor in metamorphism. The example to be described here is the contact aureole surrounding the Beinn an Dubhaich Granite, Isle of Skye,

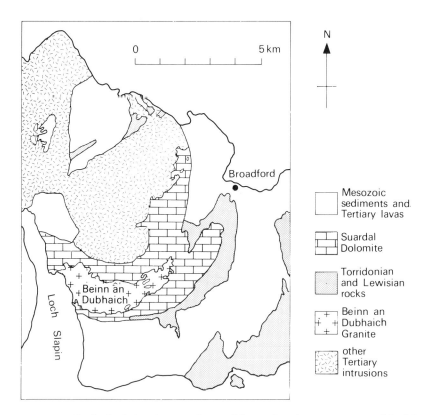

Figure 5.11 Geological sketch map of part of the tertiary igneous centre of the Isle of Skye, showing the Beinn an Dubhaich Granite. From Stewart (1965).

Scotland (Fig. 5.11). The contact metamorphic aureole of this granite was also described by Tilley (1951).

The country rock surrounding the granite is the Suardal Dolomite. It is of Cambrian age, equivalent to the Durness Limestone further north in Scotland. It is a dolomite rock or dolostone containing oval nodules of chert, often concentrated in layers parallel to the bedding. Close to the contact of the granite, the dolomite of the country rock has been altered by contact metamorphism. The $MgCO_3$ component has broken down by the following chemical reaction:

$$\underset{\text{dolomite}}{CaMg(CO_3)_2} \rightleftharpoons \underset{\text{calcite}}{CaCO_3} + \underset{\text{periclase}}{MgO} + \underset{\text{vapour}}{CO_2}$$

The CO_2, carbon dioxide gas, has escaped from the rock. This reaction was described in metamorphic rocks by the great pioneer of metamorphic petrology A. Harker, who named it **dedolomitisation**. Periclase is not stable in rocks near the surface of the ground on a Scottish hillside and has become hydrated by ground water to brucite $(Mg(OH)_2)$. The reader may wonder why the equivalent reaction in calcium carbonate has not occurred in the Beinn an Dubhaich Aureole. It is:

$$\underset{\text{calcite}}{CaCO_3} \rightarrow \underset{\text{quicklime}}{CaO} + \underset{\text{vapour}}{CO_2}$$

This reaction is used in the manufacture of quicklime, and is easily performed in the laboratory by heating powdered limestone over a gas burner. The reason the reaction is not seen in contact metamorphic rocks is that a very small amount of pressure inhibits the escape of the carbon dioxide, even at the temperatures of metamorphism. The reaction is only known in nature from limestone blocks which have fallen onto the surface of lava flows. The reaction is a reversible one and should be written:

$$CaCO_3 \rightleftharpoons CaO + CO_2$$

Most of the reactions to be discussed in this book are reversible and this way of writing them will be used from now on. A small increase in pressure drives this reaction to the left. The dedolomitisation reaction, although it also involves the loss of CO_2 from the rock, is not so strongly inhibited by increasing pressure and therefore does occur in contact aureoles.

The chert nodules introduce another chemical component, SiO_2, into the dolomite rock. As the contact with the granite is approached, reaction rims appear between the nodules and the dolomite. In the outer parts of the aureole these consist of a shell of talc, while nearer the contact remarkable alternating layers of diopside and forsterite occur. The rims formed by reactions which occurred at the surface of the nodules; but the reason for

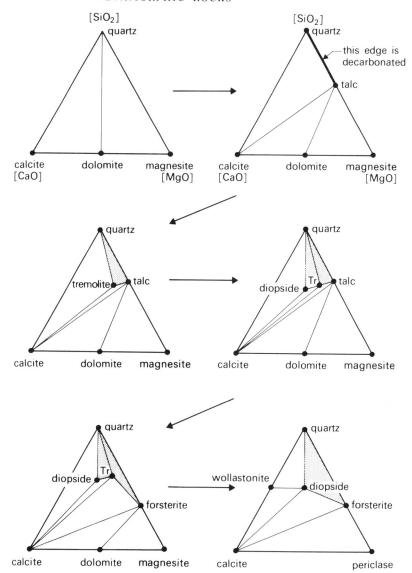

Figure 5.12 A sequence of $[SiO_2] - [CaO] - [MgO]$ triangular diagrams to illustrate the progressive decarbonation of siliceous dolomites in the Beinn an Dubhaich Aureole. Shaded parts of the triangles represent assemblages which are completely decarbonated.

the alternation is not known. Dolomite and chert have ceased to be stable together. SiO_2 has diffused out of the nodules into the surrounding dolomite, MgO into the chert to form the talc rims. Closer to the contacts of the granite, CaO has also diffused across the boundary.

In the reaction rims, a sequence of new minerals appears with increasing metamorphic grade. From low to high grade the sequence is: talc, tremolite, diopside, forsterite, periclase, wollastonite. They appear usually as individual reaction rims surrounding the chert nodules, except for the periclase which occurs in the dolomite rock, as explained above.

The changes in possible stable mineral assemblages with increasing metamorphic grade are summarised in Figure 5.12. The dolomite rock with its chert nodules can be regarded as a four component system SiO_2–CaO–MgO–CO_2. The rock compositions can therefore be represented in a tetrahedron with these four components at its corners. The triangles of Figure 5.12 are projections of points within the tetrahedron onto its $[SiO_2] - [MgO] - [CaO]$ face. The presence at low grades of the hydrous minerals talc and tremolite introduces a complication, which is overcome by neglecting the H_2O contents in plotting the diagrams.

The changes occurring in these assemblages with increasing metamorphic grade are chemical reactions involving the driving out of the rock of CO_2. Thus the sequence may be described as showing **progressive decarbonation**. A simple reaction occurring in the sequence is the reaction of chert with calcite to form wollastonite, as follows:

$$\underset{\text{calcite}}{CaCO_3} + SiO_2 \rightleftharpoons \underset{\text{wollastonite}}{CaSiO_3} + CO_2$$

The carbon dioxide has escaped from the rocks during metamorphism, like the water from the pelitic rocks of the Skiddaw and Comrie Aureoles. An indication of this process is given in Figure 5.12, where the parts of the composition triangle over which the mineral assemblages include no carbonate are indicated by shading and heavier tie-lines. These assemblages have been completely decarbonated by the contact metamorphism.

The process of decarbonation does not always end with the highest-grade assemblages shown in Figure 5.12. These assemblages are, however, of the highest grade found in ordinary contact metamorphism. In a few contact aureoles unusual geological circumstances have caused unusually intense heating of the country rocks and decarbonation has gone further, although the calcite in pure calcite rocks has still not been decarbonated. The most famous of these unusual contact aureoles is round a dolerite plug in chalk at Scawt Hill, County Antrim, Northern Ireland. The unusually high-grade decarbonation has produced a number of rare and interesting calcium and magnesium silicate minerals. Descriptions of these, and of the process of progressive decarbonation forming them, will be found in Winkler (1976; pp. 133–37) and Miyashiro (1973, pp. 291–92).

The low-grade hydrous minerals of the Beinn an Dubhaich contact aureole have been mentioned. To form talc and tremolite from chert and dolomite required the introduction of H_2O into the rock. Obviously, detailed consideration of the process of contact metamorphism should include the introduction of H_2O at low grades, and the driving out of both CO_2 and H_2O at higher grades. This will be generally true for the metamorphism of rocks containing both hydrous and carbonate minerals. The processes of dehydration and decarbonation with increasing temperature which occur in metamorphosed carbonate-bearing sediments resemble processes which are important in the cement and ceramic industries and so their chemical nature is quite well understood. The subject is discussed at length in Winkler (1976, Ch. 9), and a specific example of successive decarbonation and rehydration described by Agrell (1965).

It is important to appreciate that the sequence of new minerals which appear with increasing grade, and of old minerals which disappear, in the pelitic rocks of the Skiddaw and Comrie Aureoles and in the carbonate rocks of the Beinn an Dubhaich Aureole, depend partly upon the composition of the fluid which was driven off during metamorphism. For Skiddaw and Comrie this fluid was rich in H_2O, for Beinn an Dubhaich it was rich in CO_2. The investigation of the composition of fluids present during metamorphism is an active field of research at the moment and may have important results for the interpretation of both contact and regional progressive metamorphic sequences. Already some older interpretations of the conditions of metamorphism of rocks in terms of temperature and pressure alone have been replaced by interpretations considering temperature, pressure and fluid composition.

Exercise

Here are four analyses of rocks from the Skiddaw Aureole. Plot their compositions onto two AFM triangular diagrams. On one of these, sketch tie-lines representing the mineral assemblages of the Skiddaw Slates unaffected by contact metamorphism; on the other, sketch tie-lines representing mineral assemblages of the inner hornfels zone of the contact aureole. Take the general composition ranges of the minerals from Figure 5.5, sketch the tie-lines, and compare your result with the Figure.

	(1)	(2)	(3)	(4)	
SiO_2	58·41	55·25	63·18	54·37	
TiO_2	1·00	1·12	0·98	1·03	
Al_2O_3	20·25	22·38	19·29	22·11	
Fe_2O_3	0·63	0·72	0·27	2·23	
FeO	8·05	8·59	6·55	6·52	
MnO	0·07	0·54	0·10	0·36	
MgO	2·02	2·06	1·86	2·04	
CaO	0·41	0·82	0·46	0·46	
Na_2O	0·68	1·08	1·04	0·73	
K_2O	2·50	2·70	3·81	4·94	
H_2O+	4·87	3·49	2·12	4·42	
H_2O-	0·46	0·50	0·19	0·15	n.d. = not
P_2O_5	0·23	0·13	0·11	0·00	determined
BaO	0·04	0·03	0·05	0·05	
C	0·39	0·13	n.d.	n.d.	
Others	n.d.	0·69	0·11	0·96	
Total	100·01	100·23	100·12	100·37	

(Analyses from Eastwood *et al.* (1968), reproduced by permission.)

Note Since these rocks contain no feldspar, it must be assumed that the small proportion of sodium is substituting for potassium in micas, especially muscovite. Therefore $3 \times [Na_2O]$, as well as $3 \times [K_2O]$, must be subtracted from $[Al_2O_3]$ to obtain a value for A.

6

Dynamic metamorphic rocks

The local occurrence of dynamic metamorphic rocks was emphasised in Chapter 2. Although individual **mylonites** have often been studied with the petrological microscope, descriptions of several samples in a progressive metamorphic sequence are rarer. In this chapter, two such sequences will be discussed, one illustrating dynamic metamorphism associated with thrusting, the other shock metamorphism at a meteorite impact site. In both cases the country rocks which have been altered are granites, using that rock name in a wide sense. Brittle rocks such as granite and basic igneous rocks more often show dynamic metamorphic effects than weaker pelitic rocks. The reason for describing granites rather than basic igneous rocks in this chapter is that the examples are familiar to the author, whereas comparable examples of progressive dynamic metamorphism in basic igneous rocks are not.

A feature of dynamic metamorphic rocks is that, because mechanical processes of deformation and fragmentation are distinctive to their metamorphism, textures are more important, and lists of minerals present less important, than in contact or regional metamorphic rocks. Many dynamic metamorphic rocks do not contain equilibrium assemblages of minerals. Some do, and these are of interest because conditions of temperature, pressure and fluid composition during deformation may be estimated from them. The importance of such results for tectonic studies is obvious and one famous mylonite will be described to illustrate this point.

PROGRESSIVE DYNAMIC METAMORPHISM OF GRANITE

The example to be described here is from south-east Turkey, near the city of Bitlis (Fig. 6.1a). This progressive dynamic metamorphic sequence has been described by Hall in an unpublished thesis, and permission to use his unpublished results is gratefully acknowledged. This part of Turkey lies in the southern branch of the Alpine–Himalayan mountain chain (Fig. 6.1a). There is geological evidence for orogenic deformation as recently as the

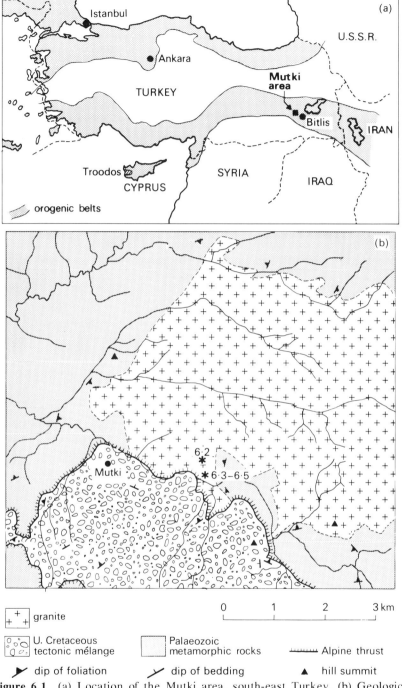

Figure 6.1 (a) Location of the Mutki area, south-east Turkey. (b) Geological sketch map of the Mutki area. Location of specimens shown in Figures 6.2–6.5 indicated.

early Pliocene (Hall 1976) and the strong seismic activity of the area suggests it may be continuing at present. The rocks round Bitlis are regional metamorphic rocks intruded by granites. The regional metamorphic rocks are of Lower Palaeozoic or perhaps even Precambrian age, the granites probably Hercynian. Both granites and country rocks have been broken up by post-metamorphic thrusts and faults. Some of the larger thrusts bring Cretaceous and early Tertiary rocks alongside the Palaeozoic rocks (Hall 1976, Mason 1975). In the area studied in detail by Hall, round the village of Mutki west of Bitlis, an intrusive granite is in contact with one such thrust on its northern side, with fragmented Cretaceous rocks to the south (Fig. 6.1b). The thrust surface is irregular in shape, but generally dips to the north with the granite above and the fragmented Cretaceous rocks below.

As the southern thrust contact of the granite is approached, dynamic metamorphism becomes more marked. This is mainly seen in the field by a change from massive, medium-grained granite to foliated granitic gneiss. There are also fault and thrust zones within the mass of the granite near which it has been altered to foliated granitic gneiss. In the most intensely altered parts, the granite has altered to fine-grained, strongly foliated mylonite. Thus the field observation of the progressive metamorphic sequence may be summarised granite → granitic gneiss → mylonite.

Although the granite away from the southern thrust appears in hand-specimen to be unaffected by dynamic metamorphism, thin section study shows that this is not the case. Figure 6.2 shows the most unaltered granite

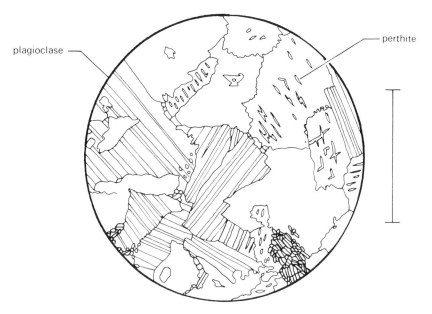

Figure 6.2 Slightly metamorphosed granite. Scale bar 1 mm.

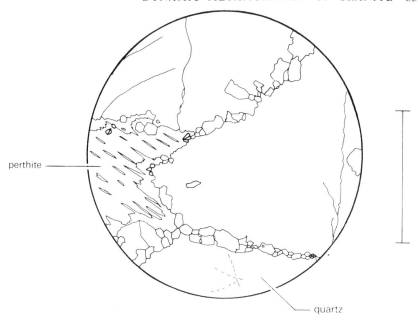

Figure 6.3 Granite with mortar texture. Scale bar 1 mm.

discovered. It consists of approximately equal proportions of potash feld-spar, plagioclase feldspar (An$_{23}$) and quartz, with a minor proportion of blue–green hornblende. These proportions of minerals make it an adamel-lite according to the strict mineralogical classification of granitic rocks. The potash feldspar is perthite, consisting of orthoclase with platy or ribbon-shaped inclusions of untwinned plagioclase. The quartz crystals show marked strain shadowing of their extinction between crossed polars, and the boundaries between quartz crystals have a lobate form. The hornblende forms fine-grained aggregates which are apparently **pseudomorphs** after larger primary hornblende crystals. The plagioclase crystals have bent twin lamellae. The hand-specimen shows that the rock is cut by fractures, which are not visible in Figure 6.2.

Figure 6.3 shows a coarser-grained granite, again with little evidence for dynamic metamorphism in hand-specimen. In this specimen the grain boundaries between quartz crystals are again often lobate and in many places small new crystals of quartz are developed both between quartz and quartz and between quartz and feldspar. This granulation of grain boun-daries is quite a common feature of deformed massive rocks and is known as **mortar texture**. The small grains are not strained and often show 120° triple junctions. Thus the grain boundaries appear to have been fragmented and recrystallised, removing the strain shadows from the small quartz grains and producing the 120° triple junctions.

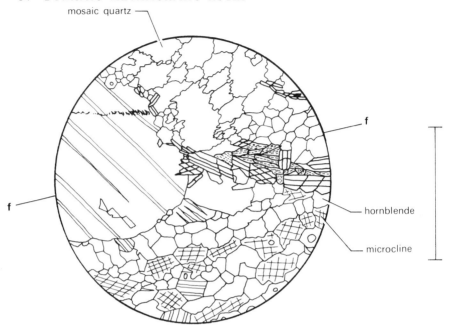

Figure 6.4 Granitic mylonite. Scale bar 1 mm. **f–f** is the foliation plane.

Figure 6.4 shows a specimen of mylonite. A plagioclase **porphyroclast** is on the left, and the foliation direction is indicated by **f–f**. In the lower half of the drawing is a strip of granular quartz crystals between the plagioclase porphyroclast and a granoblastic aggregate of quartz and potash feldspar crystals. Whereas the potash feldspar of the primary granite is orthoclase perthite, this recrystallised potash feldspar is microcline. There are no inclusions of albite. The plagioclase porphyroclast is strained and the quartz crystals in the granoblastic groundmass also show strain shadowing. The quartz at the top of the drawing consists of many crystals with lobate boundaries. The extinction positions of these crystals differ by angles of only up to 5°, and they are probably subgrains of one primary quartz crystal. Parts of the crystal were slightly rotated relative to one another during deformation. At first they may have had gradational boundaries, as in strain shadowed quartz, or in some cases there may have been fractures between them. As deformation and annealing proceeded they developed into separate grains with lobate boundaries. The striking feature of this mylonite is the variety of grain sizes, grain boundary types and grain shapes seen in one small part of the thin section. This textural heterogeneity is a typical feature of dynamic metamorphic rocks.

Figure 6.5 shows another mylonite with a rather higher proportion of matrix to porphyroclasts. In the lower centre there is a porphyroclast of plagioclase. It has lobate boundaries which appear to have developed by the partial replacement of the plagioclase by the surrounding crystals of untwinned plagioclase and quartz. Where there is a small hornblende crystal at the boundary it has been fixed, and between such crystals the lobes in the boundary bulge into the porphyroclast. This illustrates one way in which the lobate boundaries may have formed. It is possible that the grain boundaries in quartz and feldspar crystals might be fixed by concentrations of crystal defects as well as by small grains of different minerals at the grain boundary. The plagioclase porphyroclast has partly altered to clinozoisite. It appears that in this mylonite the original igneous minerals of the granite are breaking down to the metamorphic assemblage albite + quartz + clinozoisite + hornblende + opaques. The rock has a different composition from those in Figures 6.2–6.4 because potash feldspar is absent. Like the mylonite of Figure 6.4, its textural heterogeneity indicates that it has undergone dynamic metamorphism.

The evidence for recrystallisation in these rocks and the possibility that the last rock may show an approach to an equilibrium mineral assemblage of comparatively low-grade regional metamorphism, show that although mechanical fragmentation has been important in metamorphism, recrystallisation has also occurred. The break-up of large mineral grains by the development of mosaic texture, followed by recrystallisation, is also

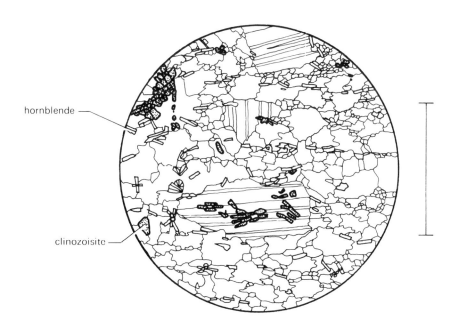

Figure 6.5   Most strongly metamorphosed granitic mylonite. Scale bar 1 mm.

important. The formation of the mylonites seems to have been complex and to have involved both fragmentation and annealing. To gain some understanding of the processes involved, detailed structural investigations in the area round the thrust or fault must be combined with examinations of the petrography of the mylonites. In the case of the Mutki granitic mylonites, such investigations indicate that the fragmentation of the Cretaceous rocks south of the thrust occurred in at least two distinct episodes, of Upper Cretaceous and Miocene–Pliocene age respectively. It is possible that dynamic metamorphism occurred in the granite during both phases.

THE LOCHSEITEN MYLONITE, SWITZERLAND

This is an example where the tectonic evolution of a thrust plane is unusually well understood and where mylonite formation is thought to be crucial to the development of the thrust. A detailed account has recently been given by Schmid (1975).

The Lochseiten Mylonite is a layer of marble 1–2 m thick which lies at the base of the Glarus Nappe in the Alps to the south of Zürich, Switzerland. This is a classic area in which Escher von der Linthe, Bertrand and Heim established for the first time the existence of large-scale overthrusting in an orogenic belt. The reader who does not know the story will find it recounted by Bailey (1935). The mylonite lies between the lowest **nappe** of this part of

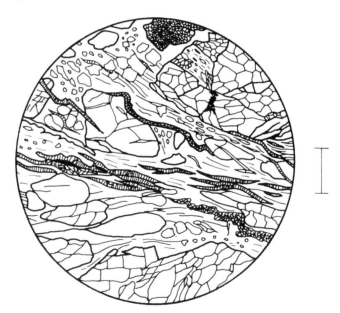

Figure 6.6 Lochseiten Mylonite, Swanden, Canton Glarus, Switzerland. Scale bar 1 mm.

the Alpine chain, the Glarus Nappe, and the untransported tectonic basement, or **autochthon**. At the time of thrusting the nappe was a comparatively rigid slab of rock about 2·5 km thick, mainly made up of a thick
Permian conglomerate. The autochthon was also rigid, formed of early
Tertiary rocks which had previously been metamorphosed to slate (Chapter 10). The mylonite is a thin layer between the two which apparently acted
as the lubricant of the thrust plane. The Glarus Nappe travelled at least 35
km over the thrust plane.

Figure 6.6 shows a thin section of the mylonite. The rock has a well-
marked but uneven foliation parallel to the plane of the thrust. It has a
typical mylonite texture with porphyroclasts of marble in a matrix of calcite
crystals too fine-grained to represent in the sketch. The calcite crystals of
the porphyroclasts are strained. Many of the porphyroclasts are recognisable fragments of calcite veins, cross-cutting the fragmental texture of
earlier formed mylonite. These veins were later fragmented themselves.
Some specimens of the mylonite consist entirely of more or less fragmented
veins of calcite in the very fine-grained matrix. The matrix grains do not
show strain shadows.

Figure 6.7 shows a generalised vertical cross section through the mylonite zone. From a comprehensive survey of the minor structures near the

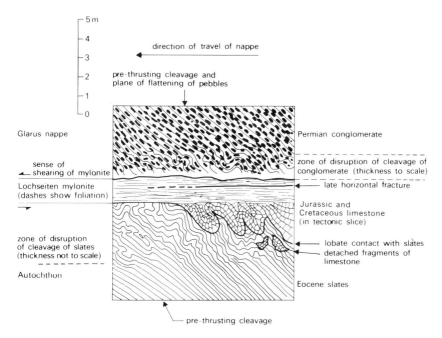

Figure 6.7   Generalised cross section through the Lochseiten Mylonite zone at the
base of the Glarus Nappe, from Schmid (1975).

thrust zone Schmid concluded that the majority of the relative movement between the nappe and the autochthon was taken up in the mylonite zone. The deformation of the parts of the conglomerate above the thrust and the slate below, and the fault movement on the late flat-lying fracture (Figure 6.7) can only account for a small proportion of the 35 km displacement. The mylonite deformed by laminar flow, during which the foliation developed. It smeared out like butter over bread from small slices of Jurassic and Cretaceous limestone which were over-ridden by the thrust (a part of one such slice is shown in Figure 6.7). The presence of the broken calcite veins suggests that the deformation of the mylonite was assisted by solution of calcite in intergranular fluid and its subsequent reprecipitation in veins. The granoblastic matrix indicates that recrystallisation rather than mechanical deformation of grains was the dominant flow mechanism.

From a study of the magnesium content of the calcite in the mylonite, Schmid argued that the maximum temperature during laminar flow was about 390 °C. This determination is based on the kind of data about mineral stability discussed in Chapter 13 of this book and assumes that the matrix of the mylonite is an equilibrium mineral assemblage. From geological evidence Schmid concluded that the minimum possible rate of travel of the Glarus Nappe over its thrust plane was 3·5 mm/year. Experiments in which marble was deformed at temperatures and pressures comparable with those indicated by the mineral assemblage of the Lochseiten Mylonite show that marble does deform as a ductile solid under these conditions. This mechanism for nappe transport is different from the one generally accepted by geologists studying the Alpine nappes in recent years. It should be emphasised that the thoroughness of both his petrographic and his structural studies is essential to Schmid's argument. If his petrological or structural conclusions are seriously wrong, the numerical assumptions

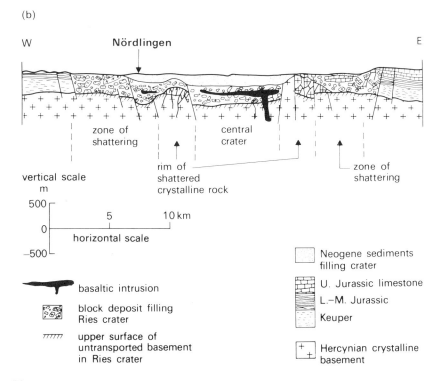

Figure 6.8   (a) Location of the Ries Crater. (b) Cross section through the Ries Crater, from Dorn (1960).

about the thickness of the flowing layer and the temperature and pressure of the mylonite are changed and the tectonic conclusions change also. Studies of this kind offer an opportunity to investigate tectonic mechanisms at least semi-quantitatively in crustal rocks, and could provide very valuable data for plate tectonic models. It is significant that a number of research teams are now studying thrusts related to subduction zones with this end in view.

PROGRESSIVE SHOCK METAMORPHISM AT THE NÖRDLINGER RIES CRATER, GERMANY

The Ries Crater is a remarkable, approximately circular hollow 21 – 24 km in diameter. It lies near the western border of the state of Bavaria in southern Germany, between the Franconian and Schwäbian hills (Fig. 6.8a). The ancient town of Nördlingen stands on Neogene sediments which fill the hollow.

The origin of the structure has been the topic of considerable discussion, now largely resolved. It is generally considered to be a somewhat eroded crater, originally formed by an immense explosion following the impact of a large meteorite upon the Earth's surface. Figure 6.8b shows an east–west cross section through the structure. The local geological succession is relatively simple. Flat-lying Mesozoic sediments unconformably overlie a complex of metamorphic and igneous rocks, which formed during the Hercynian orogeny. The Hercynian basement complex consists mainly of biotite granite containing dykes of amphibolite. The Mesozoic sediments include Keuper marls, Liassic clays and a thick limestone of Upper Jurassic age. The succession within the crater is very different (Fig. 6.8b). Beneath Neogene sediments is a remarkable deposit which consists of a jumble of blocks of biotite granite and all the sediments found above it. The block deposit rests directly on the crystalline basement complex, with a strongly discordant surface of contact between the two.

The meteorite impact origin of the Ries Crater was demonstrated by petrographic studies upon crystalline basement rocks. In the hills surrounding the crater, a sheet of material ejected during its formation is found in the Neogene sedimentary succession. This fixes the age of formation somewhere between the beginning and the middle of Upper Miocene times. The ejecta sheet consists of a tuff-like rock called **suevite**, along with larger ejected blocks of all the country rocks. The ejected blocks of biotite granite show a complete gradation between relatively unaltered granite and glassy rocks. This is interpreted as a progressive shock metamorphic sequence (Stöffler 1966). The stages in this sequence have not been documented by study of rocks from the crystalline basement beneath the crater because as a glance at Figure 6.8 will show, the collection of such rocks would involve a considerable amount of deep drilling.

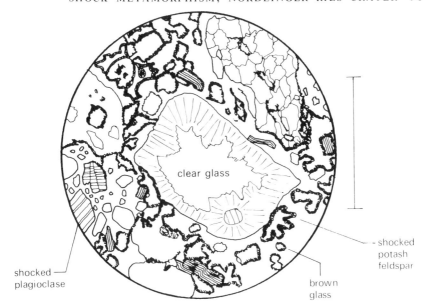

Figure 6.9 Suevite from the Ries Crater. Scale bar 1 mm.

Figure 6.9 shows a thin section of a fine-grained specimen of suevite. The matrix material is pale brown glass. This has numerous vesicle-like holes lined with darker brown glass and contains fragments of several different kinds. The fragment at the upper right is of marble, transformed from Jurassic limestone. Other fragments are of clear glass, potash feldspar, plagioclase feldspar, biotite and quartz. On any view of the mode of formation of the Ries Crater, the suevite should strictly be regarded as an extremely unusual type of sedimentary rock, composed of fragments ejected from the crater, in a glass matrix. The fragments within the rock display textural and mineralogical changes that indicate they have undergone shock metamorphism.

In this specimen, the feldspar fragments are especially illuminating. Originally the clast in the middle was probably composed entirely of potash feldspar. Its interior is of colourless isotropic glass, and its outer part is of radiating crystalline aggregates of potash feldspar. These crystals of potash feldspar appear to have crystallised from the glass. The R.I. of the glass is *higher* than that of the crystalline potash feldspar, which indicates that it is not a glass formed by melting. It is a glass-like material formed by shock waves shaking the constituent ions of the potash feldspar out of their regular positions in the crystal structure. This process can be carried out in the laboratory. Only shock waves formed by a very powerful and short-lived explosion are intense enough to do this. Artificial nuclear explosions are powerful enough, and so are meteorite impacts, but volcanic explosions are probably not. Another interesting clast is seen on the left of Figure 6.9. This is of plagioclase feldspar. The twin lamellae have been displaced by shock metamorphism into three cross-cutting zones separated by planar fracture surfaces. Biotite clasts in the drawing have their cleavage planes bent into kink bands. These structures in clasts in the suevite match those produced in granite by artificial shock waves. Although some of them, such as kink banding in biotites, may also occur by tectonic dynamic metamorphism the range of changes seen can only be satisfactorily explained by shock metamorphism.

From a study of clasts from many suevite specimens, and of larger ejected blocks of granite, Stöffler (1966) distinguished four stages of shock metamorphism, as follows:

low grade	I	Development of planar cleavage cracks in quartz.
	II	Partial-to-total transformation of feldspars to
↓		glass.
	III	Reduction of pleochroism and birefringence in biotite.
high grade	IV	Shock melting of all minerals in the granite.

In Figure 6.9 the feldspar clasts have been shock metamorphosed to stage II. The biotite clasts, although deformed, retain their pleochroism and birefringence, suggesting metamorphism below stage III. But the clast at the upper left is of clear glass containing vesicles and probably represents quartz or feldspar metamorphosed to stage III or IV. Thus one specimen of suevite contains fragments of different metamorphic grade, indicating that metamorphism occurred before the fragments were incorporated into it.

Figure 6.10 shows a thin section from an ejected granite block. The quartz and feldspars have been totally transformed to clear glass, which has a pattern of cracks like the perlitic cracks in volcanic glass. Veins of glass formed by melting of the granite to a liquid during intense shock metamorphism traverse the slide. In plane-polarised light (as the section is drawn) the parts without glass veins look like granite under low magnification, because the texture of the parent granite is retained; but under crossed polars all the grains except the biotite and opaques are seen to be isotropic glass. The biotites show kink banding, but have the normal pleochroism and birefringence. The rock is therefore from the high-grade part of the stage II.

The effects of shock metamorphism are so distinctive in thin section that there can be no doubt that the fragments thrown out of the Ries Crater underwent metamorphism by shock waves produced in an extremely powerful and very brief explosion. Most scientists working on this subject

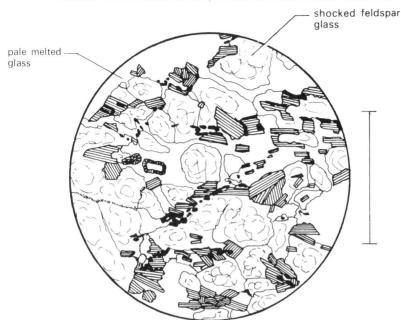

Figure 6.10 Shock metamorphosed granite from ejected block, Ries Crater. Scale bar 1 mm.

consider that explosion following meteorite impact is the only natural phenomenon which fits the case.

METAMORPHISM IN MOON ROCKS

Since the Moon has no atmosphere, meteorites strike its surface at high velocities (8–12 $km s^{-1}$). Local shock metamorphism is therefore very common in Moon rocks although the meteorites themselves have not been found because small ones are totally vapourised on impact. Matter from them has only been recognised by careful geochemical and mineralogical study of the loose lunar 'soil', more correctly called **regolith**. Since rocks of granitic composition, even in a broad sense, are extremely rare on the Moon, direct comparison of shock metamorphosed Moon rocks with the Ries Crater rocks is not possible.

The highland regions of the Moon are composed entirely of breccia. This discovery was made by the manned lunar landings Apollo 14 to 17 and was unexpected. Volcanic rocks had been widely predicted in the highlands from photogeological study but were not found. The breccias resemble the suevite and block deposit of the Ries Crater texturally, but are very different in their compositions. The fragments in the breccias are of great

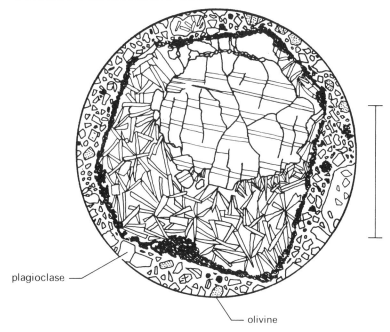

Figure 6.11 Light-coloured highland breccia from the Moon. Apollo 16 sample 60016, 80 (thin section). Scale bar 1 mm.

petrological and geochemical interest and include the oldest rocks found on the Moon, which are at least 600 Ma older than the oldest known terrestrial rocks (Taylor 1975).

Figure 6.11 shows a thin section of breccia from the Descartes Highlands region of the Moon, collected from close to the landing vehicle of the Apollo 16 mission (Warner *et al.* 1973). The author is very grateful to Dr S. O. Agrell for permission to study this thin section. The rock is a light-coloured breccia whose fragments are plagioclase crystals and fragments of anorthosite rock. The matrix is also fragmental, formed of fragments of plagioclase and olivine. The fragment which occupies most of the drawing was apparently originally a single crystal of anorthite plagioclase. The drawing shows that there is a relict of this original plagioclase in the core of the fragment. It is cut by numerous cracks, which considerably offset the twin lamellae (c.f. Fig. 6.9). The larger cracks contain isotropic shock formed glass. The outer parts of the clast consist of glass which became liquid during shock metamorphism. Evidence for this is the dusty coating on the outside of the clast, which was collected while it was still molten. The interlocking laths of plagioclase in the glass crystallised while it was still liquid. The clast was apparently incorporated into the breccia when it had solidified.

Like the Ries Crater suevite, the lunar breccia is strictly a sedimentary rock with clasts which show the effects of intense shock metamorphism. Because there are so many craters in the lunar highlands, the formation of this particular lunar breccia cannot be associated with a particular crater-forming explosion. Some breccias contain breccia clasts, presumably derived from earlier explosions, although the specimen shown in Figure 6.11 does not show this.

Because of the lack of certainty in field relations, the origin of lunar breccias is more controversial than that of the Ries Crater suevite. The majority of workers regard them as the product of meteorite impact explosions but, because the Moon has no atmosphere, it is possible that volcanic explosions may have been more intense than terrestrial ones. The ultimate origin of the anorthosite, troctolite and norite clasts in the breccias is thought to have been in layered intrusions deep beneath the Moon's crust, early in its history. Solid rocks from these intrusions were brought to the surface by immense explosions caused by the impact of bodies of asteroidal size. They may have been reworked by impacts of smaller meteorites several times since.

Shock metamorphism is not the only type of metamorphism seen in Moon rocks. Some breccias have undergone progessive recrystallisation of clasts and matrix since their formation. This can be seen in the devitrification of glass, the loss of inert gases formed by cosmic ray bombardment, the disappearance by annealing of fission tracks in minerals, and a progressive development of granoblastic texture (Warner 1972). It was first demonstrated in rocks collected from the slopes of Cone Crater during the Apollo 14 mission. These rocks probably come from different depths in the Fra Mauro formation, which is the sheet of breccia ejected by the immense explosion which formed the Imbrium Basin. Warner suggested that rocks deep in the sheet were strongly heated because the material which formed it was still hot when it fell after the explosion. The outer layers of the sheet were very poor conductors of heat, so the interior remained hot for long enough for metamorphic recrystallisation to occur. This took place at very high temperatures compared with most terrestrial metamorphism (greater than 800 °C). Warner described it as 'lunar thermal metamorphism'. It is *not*, of course, contact metamorphism. It is analogous to the welding of the base of terrestrial ignimbrite sheets, except that the source of the heat was more likely shock explosion than volcanism.

Exercise

The following analyses are of basic igneous rocks from the Mutki area, southeastern Turkey, shown in Figure 6.1. Analyses (1) and (2) are of basaltic lavas metamorphosed during the Alpine orogeny, (3) and (4) of dolerites from dykes probably metamorphosed in Palaeozoic times. The mineral assemblage of each rock is given below, and in brackets afterwards are minerals which are either relicts of the original igneous assemblage or products of retrograde metamorphism.

	(1)	(2)	(3)	(4)
SiO_2	48·22	50·52	50·51	48·73
TiO_2	0·81	1·49	1·58	1·51
Al_2O_3	15·99	13·22	16·38	15·55
Fe_2O_3	5·80	2·68	1·92	3·89
FeO	3·92	6·63	7·30	6·07
MnO	0·17	0·26	0·19	0·23
MgO	6·56	7·18	4·18	8·31
CaO	9·00	7·50	8·23	5·23
Na_2O	4·69	2·54	2·91	4·70
K_2O	0·16	2·99	1·10	0·19
H_2O total	4·12	3·87	4·51	4·40
P_2O_5	0·00	0·10	0·10	0·07
CO_2	0·10	0·12	0·11	·0·05
Total	99·54	99·10	99·02	98·93

Mineral assemblages
1. Albite + amphibole + chlorite + epidote + sphene (+ pyroxene).
2. Albite + quartz + amphibole + chlorite + epidote + sphene + apatite + opaques.
3. Albite + quartz + hornblende + garnet + epidote + sphene + apatite + opaques (+ chlorite).
4. Albite + quartz + hornblende + epidote + sphene + rutile + apatite + opaques (+ chlorite).

(Unpublished analyses by R. Hall, gratefully acknowledged.)

Plot analyses (1) and (2) onto one ACF diagram, (3) and (4) onto another. Sketch tie-lines on each diagram representing the mineral assemblages in each case. Take the compositions of epidote, chlorite, and garnet from Figure 8.13. The amphibole composition fields have the same shapes as in that Figure, but the maximum amount of aluminium in the amphibole is different in each case. Estimate the maximum amount of aluminium in the amphibole in each case. The reason for the difference will be explained in Chapter 8; it reflects different conditions of metamorphism in each case.

Note Remember to correct the C value for the small amounts of P_2O_5 and CO_2. Because the mineral assemblages contain sphene and rutile, *do not* subtract $[TiO_2]$ from $[FeO]$ for the F total.

7

Regional metamorphic rocks of Precambrian shield areas

The area selected as an example of Precambrian shield metamorphic rocks is the Lewisian Complex of northwestern Scotland. It has been extensively studied and its metamorphic rocks are similar to those of many other Precambrian shield areas. Although the commonest types of metamorphism in shield areas are those described in Chapters 8 and 9, a less common type of metamorphism is described in this chapter because it is seldom found outside Precambrian shields and because the conditions of this type of metamorphism are relevant to the general discussion of the origin of the oldest parts of the Precambrian shields. The origin of banded gneiss, a common rock in Precambrian shields, is also discussed.

The Lewisian Complex is a fragment of the Canadian Shield, separated from the Greenland part of the shield by the comparatively recent opening of the North Atlantic ocean. The predominant rock type is gneiss, often of granitic composition, with strong gneissose banding. Radiometric age determinations show that all the rocks of the Lewisian Complex are ancient, metamorphism having ceased 1700 Ma ago. One group of rocks giving older dates (2900–2200 Ma), the Scourian Complex, exists as relict patches between regions of rock metamorphosed later (2200–1700 Ma), the Laxfordian Complex. Some workers consider that the times of metamorphism and the geological structures are better described by three major metamorphic episodes (Bowes 1976). The first two events they recognise both affected rocks assigned here to the Scourian Complex, the older one being called the Badcallian metamorphic episode, the younger one the Inverian metamorphic episode. This chapter follows Read & Watson (1975a) in using the simpler two-fold division, but the existence of the different versions illustrates the complex evolution of these rocks and the difficulties of interpretation. Rocks of the Lewisian Complex have been remetamorphosed in the northwestern part of the Caledonian orogenic belt (Fig. 7.1). Repeated metamorphism is common in Precambrian shields.

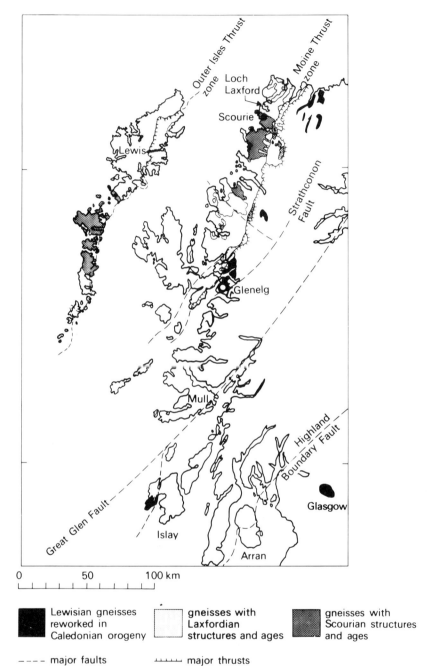

Scale bar:

0   50   100 km

Legend:
- Lewisian gneisses reworked in Caledonian orogeny
- gneisses with Laxfordian structures and ages
- gneisses with Scourian structures and ages

- - - - major faults   ┴┴┴┴┴ major thrusts

Figure 7.1 Geological sketch map of northwestern Scotland showing the Lewisian Complex, divided into areas deformed and metamorphosed during the Scourian, Laxfordian and Caledonian episodes, modified after Read and Watson (1975a).

The rocks described in this chapter come from the area around the village of Scourie in the northern part of the mainland outcrop area of the Lewisian Complex. Just north of the village, a transition from rocks of the Scourian Complex to rocks of the Laxfordian Complex is seen. The rocks are banded gneisses whose composition varies from that of acid igneous rocks to that of ultrabasic igneous rocks. The gneisses are cut by numerous dykes of basic igneous composition. The banding of the gneisses is due to variation in rock composition, the darker bands being richer in FeO and MgO, the lighter bands in SiO_2, Na_2O and K_2O. The petrology of the gneisses of the Scourian Complex of this area has been described by Sutton and Watson (1951). The petrogenesis of the gneisses has been discussed by numerous authors (e.g. O'Hara 1961, Bowes *et al*. 1964).

PETROLOGY OF PYROXENE GNEISSES

Figure 7.2 shows a thin section from a gneiss of acid igneous composition. Although the outcrop and hand-specimen show gneissose banding, the thin section is too small for it to be apparent. The mineral assemblage is quartz + feldspar + orthopyroxene + clinopyroxene + opaques. The rock also contains hornblende and biotite, but these are in reaction rims and therefore are not members of the high grade mineral assemblage listed, but

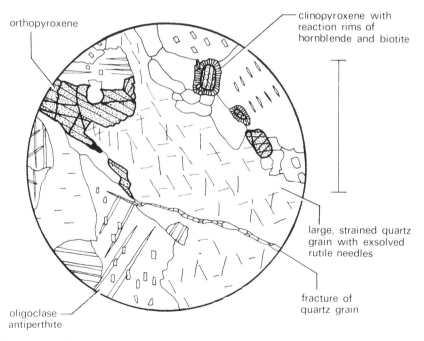

orthopyroxene

clinopyroxene with
reaction rims of
hornblende and biotite

large, strained quartz
grain with exsolved
rutile needles

fracture of
quartz grain

oligoclase
antiperthite

Figure 7.2   Charnockitic gneiss from Scourie, Sutherland. Scale bar 1 mm.

belong to a later stage of **retrograde metamorphism**. The early, high grade assemblage has a coarse-grained granoblastic texture (c.f. Fig. 5.8). The feldspar in this rock is antiperthite, crystals of oligoclase (An_{30}) containing plate-shaped inclusions of potash feldspar. Compare it with the perthite of Figure 6.2. Antiperthite forms because plagioclase accepts more potash feldspar in solid solution in its crystal structure at high temperatures than at lower temperatures. As the plagioclase cools, the potash feldspar **exsolves**, so that what was a homogeneous crystal at high temperature becomes an intergrowth of two mineral species. This will only occur if the rate of cooling is sufficiently slow for diffusion of potassium and sodium ions through the plagioclase crystal to be effective in separating the two minerals. Thus, the presence of antiperthite in the Scourie gneisses suggests that the original crystallisation of the plagioclase was at high temperature, and that it subsequently cooled slowly. The twin lamellae of the antiperthite plagioclase are slightly bent, and strain shadows also occur in quartz. This suggests that there was a certain amount of deformation after the crystallisation of the main mineral assemblage.

One interesting feature of this rock is the presence of many small needles of a mineral with a higher R.I. than the quartz as inclusions in the quartz crystals. The needles point in three regular directions, apparently parallel to symmetry axes of the quartz crystals. The needles are probably rutile (TiO_2), although they are too small for identification by the petrological microscope. The TiO_2 presumably exsolved from the quartz during cooling, like the potash feldspar from the plagioclase. The exsolved inclusions of rutile in the quartz crystals make it appear dark in hand-specimen, with a distinctive dark blue opalescence.

This rock has the chemical composition of a granodiorite, and the mineral assemblage is also like the list of minerals in a granodiorite, except that the minor amount of ferromagnesian minerals consists of anhydrous orthopyroxene and clinopyroxene, not hydrous micas or amphiboles. (The biotite and hornblende of this rock obviously crystallised later than the main mineral assemblage.) The **granoblastic** texture of the rock, and its field relationships, leave no doubt that it is a metamorphic rock.

Rocks with a granite to granodiorite composition, but with granoblastic texture and anhydrous ferromagnesian minerals, belong to the **charnockitic suite**. The name charnockite is strictly applied to the member of the suite with predominant potash feldspar (i.e. equivalent to granite in its chemical composition). The rock of Figure 7.2 with plagioclase rather than potash feldspar is strictly speaking an **enderbite** (Tilley 1936), the sodium-rich member of the charnockite suite, which contrasts with **charnockite**, the potassium-rich member.

Rocks of the charnockite suite are found in many Precambrian shield areas. The name has been given because the tombstone of Job Charnock in Calcutta, India, is made of this rock. The stone of the tomb was quarried near Madras, in southern India. There is some argument whether the Scourie rocks should be described as charnockites or not. The diagnostic feature of charnockite is the dark colour of the quartz, due to the exsolved rutile needles, as discussed earlier (Howie 1955). The problem is that the Scourie gneisses may not be dark enough to merit the name! This difficulty will be avoided here by describing the acid gneisses of Scourie as **charnockitic gneisses**, the adjective 'charnockitic' being used in a wider sense than the noun 'charnockite' (Howie, personal communication).

The gneisses at Scourie include mafic rock types. Two examples are shown in Figure 7.3 and Figure 7.4. These have approximately the chemical composition of basalt. The rock of Figure 7.3 has the mineral assemblage plagioclase (An_{46}) + orthopyroxene + clinopyroxene + opaques + apatite. The orthopyroxene and clinopyroxene have reaction rims round them, as in Figure 7.2. The clinopyroxene is surrounded by a rim of intergrown green hornblende and quartz. The opaque mineral also has a reaction rim of biotite. The plagioclase grains have lamellar twins with the same bent lamellae and strain extinction as in the charnockitic gneiss of Figure 7.2. The mineral assemblage of this rock differs little from the list of minerals in basalt. The only mineralogical difference detectable with the petrological

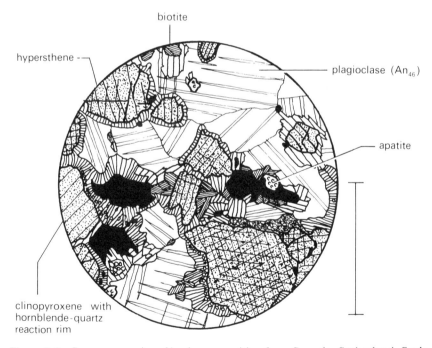

Figure 7.3 Pyroxene gneiss of basic composition from Scourie, Sutherland. Scale bar 1 mm.

microscope is the higher sodium content of the plagioclase, which is andesine rather than labradorite. The grain size is coarser than that of basalt and the texture is granoblastic rather than micro-ophitic.

Gneisses of basic igneous composition containing pyroxenes as their ferromagnesian minerals are found widely in Precambrian shield areas. There is no generally accepted name for this important rock type. The names 'charnockite' and 'enderbite' for acid igneous gneisses have already been discussed. In Saxony, Germany (German Democratic Republic), there is an area of rocks which commonly have an acid igneous composition, and mineral assemblages with anhydrous ferromagnesian minerals. They have remarkable distinctive metamorphic textures, quite different from the medium-grained granoblastic textures of typical charnockites. These rocks are called **granulites**. This name has been extended by many authors to describe gneisses of mafic composition with assemblages of anhydrous minerals, which are therefore called **basic granulites**. Other authors (e.g. Sutton and Watson 1951) extend the name charnockite in a similar way and speak of 'basic charnockites'. The problem of naming these rocks has been further complicated by the use of the name 'granulite' for a metamorphic facies (Ch. 13). Other geologists, notably the Scottish Geological Survey, used the name granulite as a descriptive term for massive rocks, rich in quartz and feldspar with a granoblastic texture,

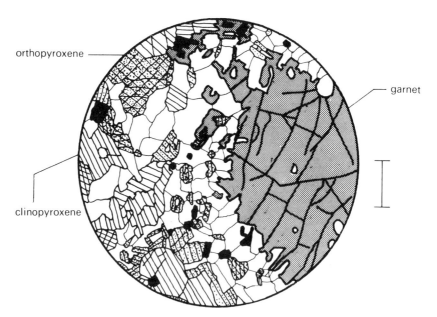

Figure 7.4 Pyroxene gneiss of basic igneous composition from Scourie, Sutherland, showing large garnet porphyroblast. Scale bar 1 mm.

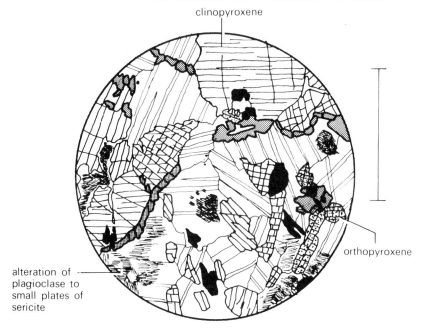

clinopyroxene

orthopyroxene

alteration of
plagioclase to
small plates of
sericite

Figure 7.5 Pyroxene gneiss of basic igneous composition from Scourie, Sutherland, showing interstitial grains of garnet. Same specimen as Figure 7.4. Scale bar 1
mm, note difference in scale.

but without the anhydrous mineral assemblages of charnockites and the
granulites from Saxony.

Because of this confusion, the name 'granulite' is probably best avoided.
In this book, rocks of basic igneous composition with gneissose banding,
anhydrous mineral assemblages and granoblastic textures, are called
pyroxene gneisses. Rocks with acid igneous compositions and anhydrous
mineral assemblages are called **charnockitic gneisses**. The name granulite is
reserved for the distinctive Saxon rocks, and for the metamorphic facies
name.

Figures 7.4 and 7.5 show another pyroxene gneiss from Scourie. This has
almandine garnet in the mineral assemblage. There are two textural varieties of garnet: large **porphyroblasts** (Fig. 7.4) and small grains in rims
between pyroxene and plagioclase (Fig. 7.5). The porphyroblasts have
very irregular outlines and numerous inclusions of plagioclase. The crystal
form reflects the mode of growth of the garnets. They apparently began by
rimming orthopyroxene grains, gradually replacing them, and finally growing outwards along plagioclase–plagioclase grain boundaries. This final
stage, showing partial engulfment and replacement of plagioclase crystals,
is visible at an arrested stage in Figure 7.4. Thus the cores of the por-

phyroblasts may include garnet which crystallised at the same time as the pyroxenes, making the early mineral assemblage of this rock plagioclase + orthopyroxene + clinopyroxene + garnet + opaques. Compared with the rock of Figure 7.3, this rock shows more extensive alteration. The plagioclase is clouded because of partial alteration to clinozoisite, white mica and more sodic plagioclase. The clinozoisite and white mica are hydrous minerals, so this alteration represents the replacement of anhydrous minerals by hydrous ones. This change implies the introduction of H_2O to the rock, during lower-grade retrograde metamorphism. In this specimen, retrograde metamorphism has not proceeded far enough to enable the retrogressive metamorphic mineral assemblage to be determined.

Figure 7.6 shows a gneiss of a different composition, that of ultrabasic igneous rock. It has the mineral assemblage clinopyroxene + orthopyroxene + olivine + spinel + pargasite + opaques. Pargasite is a palecoloured, aluminium-rich variety of hornblende. The spinel is brownishgreen **pleonaste**. The olivine has altered during retrograde metamorphism to serpentine + magnetite. This rock shows that regional metamorphic rocks, like the contact metamorphic hornfelses from Comrie discussed in Chapter 5, may have mineral assemblages undersaturated in SiO_2. It also

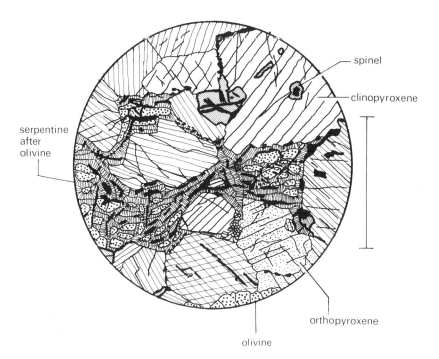

Figure 7.6 Pyroxene-pargasite-olivine gneiss of ultrabasic composition from Scourie, Sutherland. Scale bar 1 mm.

shows that in this rock composition, very rich in MgO and FeO, the hydrous mineral pargasite ($NaCa_2Mg_4Al_3Si_6O_{22}(OH)_2$) is stable in the early, high-grade mineral assemblage. Hornblende is quite often recorded in high-grade pyroxene gneisses of Precambrian shield areas. The distinctive feature which indicates the relatively high metamorphic grade of pyroxene gneisses and charnockites is the coexistence in some metamorphic mineral assemblage of orthopyroxene and clinopyroxene. If the two pyroxenes have survived from a primary igneous rock, this can be recognised by textural study, and of course does not indicate high metamorphic grade. The presence of either orthopyroxene or clinopyroxene on its own in the metamorphic assemblage also does not necessarily indicate high metamorphic grade, because either pyroxene can appear on its own in mineral assemblages of lower grades in rocks of suitable compositions.

Although these high-grade pyroxene gneisses are commonest in Precambrian shield areas, they also occur occasionally in Phanerozoic regional metamorphic areas. The mineral assemblages of these rocks are very like those of high-grade contact hornfelses and, like hornfelses, they probably underwent metamorphism at relatively high temperatures. The presence of the garnet in the mineral assemblages of basic igneous rocks suggests a difference in conditions during metamorphism from those in contact aureoles. Although pelitic rocks do not occur in the Lewisian Complex at Scourie, they are known from other areas of pyroxene gneisses such as Broken Hill, Australia (Binns 1965). They contain minerals such as sillimanite, garnet, cordierite and andalusite. A comparable type of regional metamorphic sequence in pelitic rocks of the Caledonian orogenic belt will be discussed in Chapter 9. It will be shown there that this kind of progressive regional metamorphic sequence in pelitic rocks probably formed at relatively low pressures.

Although pyroxene gneisses are found in many Precambrian shield areas, they are by no means always present. The evidence for retrograde metamorphism to mineral assemblages with hydrous minerals has been noted in some of the Scourie gneisses. The Laxfordian gneisses of the Lewisian Complex have hydrous mineral assemblages, comparable with those of regional metamorphic rocks of orogenic belts to be described in Chapters 8 and 9. The Lewisian rocks reworked in the Caledonian orogenic belt also develop hydrous, lower-grade mineral assemblages, except in a few relict blocks. The mineral assemblages of these rocks will not be described in this chapter, because accounts of comparable rocks will be given later. However, like the high grade pyroxene gneisses, and unlike the rocks described in Chapters 8 and 9, the Laxfordian and Caledonian gneisses display gneissose banding.

GNEISSOSE BANDING

At Scourie gneissose banding is found in rocks of granodioritic, basaltic and ultra-basic compositions. As mentioned earlier, it is usually too coarsely developed to be visible in thin sections, but is clearly seen in hand-specimens and field outcrops. In the Lewisian Complex, and in Precambrian shields generally, gneissose banding occurs in gneisses containing hydrous minerals such as micas and amphiboles, as well as in pyroxene gneisses. Many Lewisian gneisses are not banded gneisses but foliated granites (Table 3.1). The dark bands of banded gneisses are rich in ferromagnesian minerals—pyroxenes in pyroxene gneisses and biotite and garnet in ordinary banded gneisses. The light bands are rich in quartz and feldspar in both cases. The bands vary in thickness from 1 mm to 1 m or more (Fig. 2.8). The mode of origin of this banding has been the subject of considerable speculation and some theories will be reviewed here.

One obvious possibility is that the gneissose banding is a **relict** of the original bedding (or igneous layering in the case of metamorphosed igneous rocks) which has survived high-grade metamorphism. Gneissose banding is on a finer scale than most bedding, so if it is relict bedding the rocks must have undergone a considerable amount of tectonic flattening normal to the bedding direction. Support for an origin of gneissose banding from bedding can be found in some Lewisian gneisses. In some areas there are schists and marbles, metamorphosed during the Scourian episode, which retain structural features indicating that they are metamorphosed sediments. They may be traced laterally into rocks in which there are light-coloured bands of granodioritic composition. The granodiorite bands appear progressively parallel to the schistosity planes, which in turn are parallel to the bedding. One theory, illustrated in Figure 7.7, suggests that the granodiorite layers were offshoots of an intrusion of granodiorite, intruded along the schistosity planes. This process may be called granite injection (using the name granite in a broad sense) and gneisses formed in this way are given the name **injection gneisses**. Granite layers which cut across the banding of the gneisses are often intensely folded, supporting the suggestion made above that there has been considerable tectonic flattening.

An alternative theory suggests that the granitic material was not injected from a source of magma outside the rock, but derived by partial melting from the sediments. Layers rich in quartz and feldspar would tend to melt before those rich in mica, and it is likely that, once appreciable quantities of melt were present, the liquid might become segregated into layers among the refractory unmelted rock. This theory suggests that the rock as a whole was in a semi-molten condition with little strength at the climax of gneiss formation, and the style of the minor structures in many gneisses lends support to this idea. This mode of origin is illustrated in Figure 7.8.

These theories for the origin of the compositional banding of gneisses

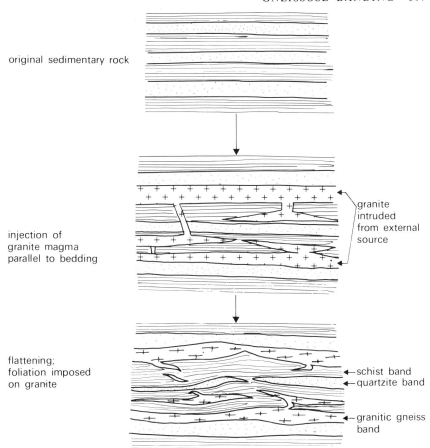

original sedimentary rock

injection of
granite magma
parallel to bedding

granite
intruded
from external
source

flattening;
foliation imposed
on granite

schist band

quartzite band

granitic gneiss
band

Figure 7.7 Cartoon to illustrate the formation of gneissose banding by injection of granite magma.

account for the interlayering of granitic or granodioritic rocks with high-grade metamorphic rocks. Rocks in which rock with a broadly granitic composition has been intermixed with high-grade metamorphic rocks are known as **migmatites**. There is a considerable variety of structural types, and their origin has been a source of much argument. It will be discussed later in this chapter. A review of the different types of migmatites and their origins has been given by Mehnert (1968).

Many gneisses have compositional banding without bands of granitic composition (e.g. the basic and ultrabasic gneisses of Scourie). It is possible that such banding may be primary igneous layering. If this is the case, the thickness of the compositional bands suggests that the primary layering was of the rhythmic type (Wager & Brown 1968) which has been flattened

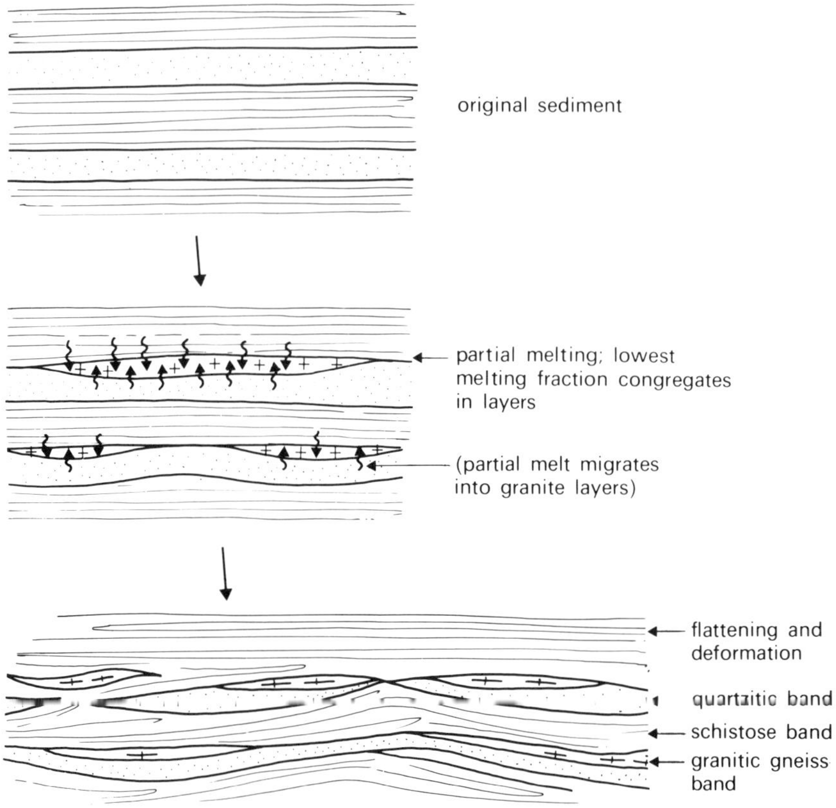

original sediment

partial melting; lowest melting fraction congregates in layers

(partial melt migrates into granite layers)

flattening and deformation

quartzitic band

schistose band

granitic gneiss band

Figure 7.8 Cartoon to illustrate the formation of gneissose banding by partial melting.

during deformation and metamorphism. This view of the origin of the banding in the basic igneous rocks has been argued by Bowes *et al*. (1964). Another possibility which might explain gneisses of many different compositions is that the banding was formed by local migration of components during metamorphism (O'Hara 1961). Potassium, sodium and silicon might have migrated into layers which are now rich in feldspars, iron and magnesium into layers which are now rich in ferromagnesian minerals. This chemical mechanism is called **metamorphic differentiation**. It is illustrated in Figure 7.9. Gneisses which have originated by this process should retain the overall composition of the primary sedimentary or igneous rocks, but individual bands could have strongly contrasting compositions which do not resemble the compositions of any known igneous or sedimentary rocks.

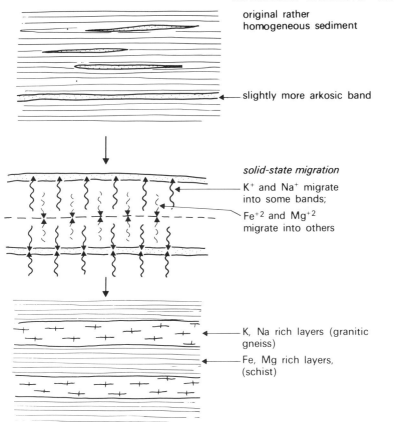

original rather
homogeneous sediment

slightly more arkosic band

solid-state migration

K$^+$ and Na$^+$ migrate
into some bands;

Fe^{+2} and Mg^{+2}
migrate into others

K, Na rich layers (granitic
gneiss)

Fe, Mg rich layers,
(schist)

Figure 7.9 Cartoon to illustrate the formation of gneissose banding by meta-morphic differentiation.

If the gneissose banding has arisen by granite injection, partial melting or metamorphic differentiation, it is not necessarily the case that the present banding is parallel to the original bedding in sediments, although for simplicity of presentation it has been shown like this in Figures 7.7–7.9. The granite veins, or bands of feldspar-rich metamorphic differentiate may cross-cut the bedding following cross-cutting structural elements such as joints, cleavage planes or fold axial surfaces. It is also possible that the formation of gneissose banding is directly related to tight-to-isoclinal folding of sediments during metamorphism. Figure 7.10 illustrates such a theory, in which mechanical segregation of material during deformation has given rise to the banding. It is based on the views of Ayrton (1969).

The theories reviewed above are not a complete list of the possibilities and they are not mutually exclusive. For example, it is very likely that the

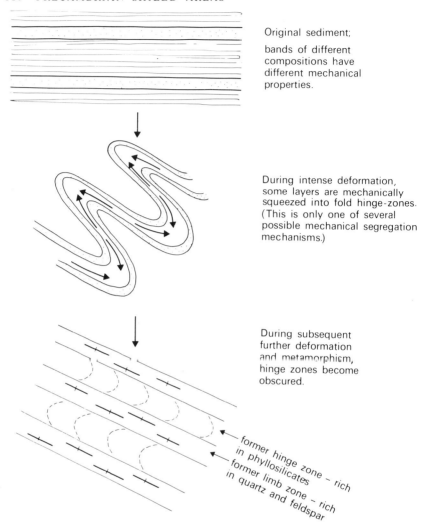

Original sediment;

bands of different compositions have different mechanical properties.

During intense deformation, some layers are mechanically squeezed into fold hinge-zones. (This is only one of several possible mechanical segregation mechanisms.)

During subsequent further deformation and metamorphism, hinge zones become obscured.

former hinge zone – rich in phyllosilicates

former limb zone – rich in quartz and feldspar

Figure 7.10   Cartoon to illustrate the formation of gneissose banding by mechanical segregation during near-isoclinal folding.

mechanical segregation of minerals into layers envisaged in the last theory could have been combined with chemical segregation of elements envisaged in metamorphic differentiation theories. In most cases, the mechanism of origin of gneissose banding cannot be determined with certainty, although the intense tectonic flattening mentioned earlier seems to be a very frequent feature. In some suitable cases the total amount of flattening, or a minimum value, may be determined from geological strain markers or by analysis of metamorphic textures.

THE ORIGIN OF MIGMATITES

The origin of migmatitic gneisses was the subject of a famous controversy in the 1940s and 1950s. Many geologists argued that migmatites had formed entirely by metamorphic processes, the granitic layers having formed by transformation of sediments in the solid state, without ever being molten. Strictly speaking, this is a type of **metasomatism**. By extension, they went on to argue that many or even all non-migmatitic granites had also formed by solid-state transformation. The most convincing evidence in support of this theory is the progressive sequence, often found in the field:

high-grade regional metamorphic rocks → migmatitic gneisses → foliated granites → massive granites

Because most sediments do not have the composition of granite, the overall composition of a very large volume of rocks must have changed, and the theory differs from those discussed in the previous section because it involves the introduction and removal of chemical components on a regional scale, i.e. **regional metasomatism**. Those who supported this view had many different ideas about the mechanism by which the addition and removal of the chemical components might have occurred. All such processes leading to the formation of granite were described under the general name **granitisation**.

These views were strongly opposed by those geologists who clung to the traditional belief that granite is an igneous rock, which has crystallised from silicate-rich melt, or magma. These geologists took a view of the nature of migmatites similar to that illustrated in Figure 7.7 and argued that the granitic parts of migmatites had been intruded as magma. The contrast in the two views is striking: in granitisation migmatites and granites are proposed to be metamorphic rocks in the sense in which they have been defined in this book; in injection they are proposed to be igneous rocks. The development of one side of the argument and the field evidence on which granitisation theories were based, can be found in Read (1957). A briefer account is given in Hatch *et al*. (1972, pp. 257–69).

In recent years the argument had died away. Probably the most important reason for this has been the increasing evidence for the origin of many migmatites by partial melting, as illustrated in Figure 7.8. The arguments for this theory are well summarised by Winkler (1976, pp. 278–324) and given at greater length by Mehnert (1968). If most migmatitic gneisses are the products of partial melting, the sequence from high-grade metamorphic rocks through migmatites to granites discussed above represents the high-temperature boundary between metamorphic and igneous rocks. Granites are igneous rocks formed by partial melting, and migmatites are truly mixed rocks comprising igneous granite mixed with high-grade metamorphic rocks. Thus there is more agreement among petrologists than there was at the height of the granite controversy, although the origin of migmatitic granites is by no means entirely settled.

The controversy is now perhaps of more interest to the historian of science than to the student trying to understand metamorphic rocks. It needed to be mentioned for two reasons. It left its mark on the description of metamorphic rocks, so that the student will only understand some accounts of regional metamorphic rocks, especially Precambrian shield rocks, if he realises that they were written under the assumption that regional metamorphic sequences represent the early stages of granitisation. It also illustrates how a considerable, integrated body of geological theory, concerned with the evolution of metamorphic rocks, may be highly regarded by one generation of geologists and yet fall into disrepute with the next. In the author's opinion this illustrates the danger of giving too much authority to theory when discussing metamorphic rocks. Because of the uncertainty of our understanding of metamorphic processes, it is quite possible that large parts of presently accepted petrogenetic theory may similarly fall out of favour. Investigations of metamorphic rocks which have had as their primary aim the validation of such a theory will then be of little value. More descriptive accounts of the rocks will retain some value because they include rock and mineral descriptions which are independent of petrogenetic theories.

Exercise

These analyses are of amphibolites from mafic bands in Precambrian gneisses of southwestern Greenland. The analyses were performed by a physical method (X-ray fluorescence analysis) and the results corrected by computer into a form in which the sum of the oxides analysed is exactly 100%, on an anhydrous basis (i.e. no H_2O value is quoted). This does not affect the *relative* values of A, C and F, so plot the analyses on an ACF diagram. The mineral assemblages are listed below.

	(1)	(2)	(3)	(4)	(5)	(6)
SiO_2	48·43	48·37	50·54	50·50	55·60	49·60
TiO_2	1·30	0·87	0·70	1·09	0·69	0·67
Al_2O_3	14·68	15·09	13·77	13·52	15·53	14·12
Fe_2O_3	3·72	3·60	3·49	3·85	2·45	3·48
FeO	9·28	8·98	8·73	9·61	6·10	8·70
MnO	0·24	0·20	0·20	0·22	0·16	0·21
MgO	5·45	8·32	8·58	7·09	6·00	8·88
CaO	13·11	11·67	10·88	11·52	9·46	11·88
Na_2O	3·02	2·46	2·58	2·39	3·32	2·04
K_2O	0·77	0·45	0·52	0·22	0·69	0·42

Mineral assemblages
1. Hornblende + plagioclase (An_{40-45}) + diopside + quartz + zircon + sphene + apatite + opaques (+ epidote).
2. Hornblende + plagioclase + diopside + quartz + biotite + apatite + a little garnet.
3. Hornblende + plagioclase + diopside + garnet + quartz + apatite.
4. Hornblende + plagioclase + diopside + garnet + quartz + apatite + opaques.
5. Hornblende + plagioclase + diopside + quartz + apatite + opaques.
6. Hornblende + plagioclase + quartz + apatite + opaques.

(Unpublished analyses by R. M. F. Preston, gratefully acknowledged.)

Taking mineral compositions from Figure 8.13 (add diopside yourself) try to sketch tie-lines to indicate the mineral assemblages. You will find that it is impossible. Why? The rocks display well developed granoblastic texture and are apparently equilibrium assemblages under the petrological microscope. (The analyses form a tight cluster on the diagram. They lie close to tie-lines from plagioclase to hornblende. However, the garnet-bearing rocks are not systematically richer in the A and F components, but appear to lie randomly among the non-garnet-bearing rocks. According to Preston, who has obtained the result on many more analyses than plotted here, the presence or absence of garnet is due to varying hydration of the amphibolites. The more H_2O-rich rocks have hydrous assemblages with plagioclase + hornblende, which may be represented on the ACF diagram; the more H_2O poor rocks have assemblages including the four phases plagioclase + garnet + diopside + hornblende, which cannot. Because H_2O is available only to a limited extent, it should be regarded as an additional chemical **component** in the rocks and the assumptions behind the ACF diagram are invalid. If the H_2O in the rock were analysed, the analyses might be plotted within an ACF–H_2O tetrahedron, and this possibility could be checked.)

8

Regional metamorphic rocks of Palaeozoic orogenic belts, I

Many classic studies of progressive regional metamorphic rocks were made in parts of the Caledonian fold belt of northwestern Europe. The work of Barrow (Ch. 2), who mapped the rocks of the Grampian Highlands of Scotland to the west of Aberdeen for the British Geological Survey during the last decade of the nineteenth century, is one such geological classic. For this reason it is thought that the reader of this book is likely to be familiar at least with the broad outlines of the progressive regional metamorphic sequence of the Scottish Highlands, so a different example will be discussed in this chapter.

Progressive regional metamorphic sequences of a similar nature to that of the Grampian Highlands are found throughout the Caledonian fold belt. The example described here is from Sulitjelma, Norway (Vogt 1927, Henley 1970). This is a copper mining field, north of the Arctic Circle and close to the border with Sweden (Fig. 8.1a). In this part of the Scandinavian Caledonides the metamorphic grade increases from south-east to north-west, and the large-scale structural pattern in the rocks at Sulitjelma means that individual stratigraphical formations may be followed from lower- into higher-grade areas. This is unusual because stratigraphical formations more commonly run parallel to the direction of the orogenic belt, which is also the direction of higher and lower grade zones of regional metamorphism. This parallelism of metamorphic zones and stratigraphical formations makes study more difficult, because it is hard to decide whether differences in the metamorphic mineral assemblages are due to changes in metamorphic grade or to changes in rock composition between the different stratigraphical units.

In Sulitjelma two stratigraphical units can be traced continuously from low- to high-grade parts of the area. They are the Furulund Schist and the Sulitjelma Amphibolites. The Furulund Schist consists of rocks of pelitic composition, whereas the Sulitjelma Amphibolites consist of

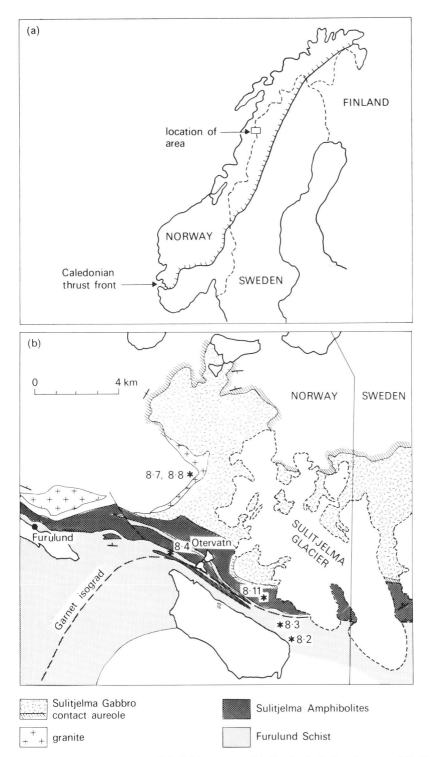

Figure 8.1 (a) Location of Sulitjelma area. (b) Geological sketch map of Sulitjelma. Locations of specimens in Figures 8.2, 8.3, 8.4, 8.7 and 8.11 shown. Isograd from Henley (1970).

metamorphosed basic igneous rocks. The progressive regional metamorphism of basic igneous rocks is also illustrated in this area by a number of small intrusions, originally of dolerite, which were intruded into the Furulund Schist before the main regional metamorphism. Near the Swedish border in the eastern part of the Sulitjelma area, there are bands of marble in the Furulund Schist containing fossils indicating an uppermost Ordovician or early Silurian age. The Sulitjelma Amphibolites overlie the Furulund Schist structurally, but probably underlie it stratigraphically.

PROGRESSIVE REGIONAL METAMORPHISM OF THE FURULUND SCHIST GROUP

The Furulund Schist consists predominantly of phyllosilicate minerals such as muscovite, biotite and chlorite. They also contain quartz and plagioclase feldspar. They contain accessory amounts of an opaque phase or phases, such as magnetite and pyrite, sphene, apatite and minerals of the epidote group. In addition, some samples may contain garnet, needle-like amphiboles and carbonate minerals (calcite or dolomite).

The original bedding of the Furulund Schist cannot be seen in hand-specimen or under the microscope. It is visible in outcrops as a darker or lighter green colouration. Local variations in rock composition giving rise

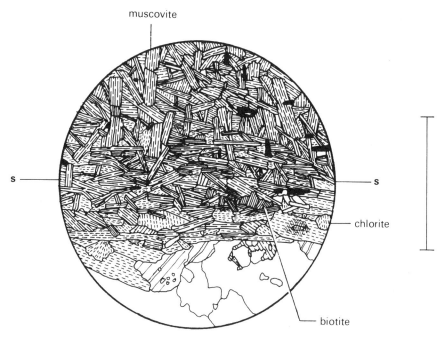

Figure 8.2   Chlorite schist. s−s schistosity direction. Scale bar 1 mm.

to distinctive mineral features, such as needle-like porphyroblasts of amphibole (Fig. 2.9), may be mapped for several hundred metres and also appear to represent the original bedding. The bedding is usually parallel to the **schistosity**, but regional studies have led to the discovery of areas where the cleavage cuts the bedding, and these have been shown to be localities at or near the hinges of folds which formed before or during metamorphism.

The progressive increase in metamorphic grade from south-east to north-west is marked by a general increase in the grain size of the metamorphic minerals, although this is often masked by considerable local variations. The phyllosilicate minerals define the schistosity, which is a penetrative planar preferred orientation (Ch. 2). Figure 8.2 shows a chlorite–biotite schist from a lower-grade part of Sulitjelma in which the character of this penetrative schistosity may be studied in more detail. There is only a statistical predominance of one preferred direction and individual phyllosilicate flakes may be oblique to the schistosity plane or even at right angles to it, although the number of such flakes is significantly less than of those which are parallel or nearly parallel to the schistosity direction.

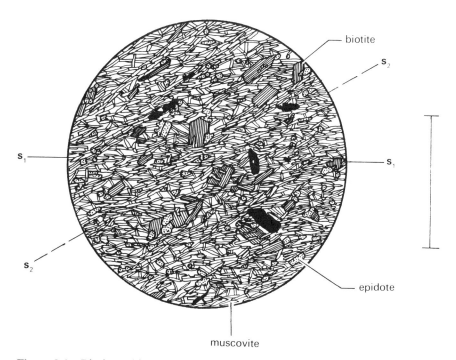

Figure 8.3   Biotite schist. $s_1–s_1$ penetrative early schistosity. $s_2–s_2$ later crenulation cleavage. Scale bar 1 mm.

In many samples of the Furulund Schist, in addition to the penetrative schistosity, there is a different, *local* orientation of the mica and chlorite grains along a number of distinct surfaces running through the rock. An example is shown in Figure 8.3, which represents a biotite schist of approximately the same metamorphic grade as the chlorite schist of Figure 8.2. The family of parallel surfaces (s_2) show a different, local orientation of the muscovite, chlorite and biotite flakes from the **penetrative** schistosity (s_1). These surfaces are said to define a **crenulation cleavage** (also known as **strain–slip** cleavage) which is a **non-penetrative** fabric because it is only developed by part of the population of phyllosilicate grains in the rock. It is crucial to the recognition of crenulation cleavage to show that it formed later than the penetrative schistosity. In the case of the biotite schist of Figure 8.3 this can be convincingly demonstrated because the muscovite and chlorite flakes defining the s_1 schistosity can be seen to swing gradually into the new orientation along the s_2 strain–slip cleavage zones. The sequence of events may be confirmed by field study, which shows that specimens of Furulund Schist with a strain–slip cleavage are confined to the hinge zones of a set of folds which fold the s_1 schistosity.

In Figure 8.3 it can also be seen that the larger biotite **porphyroblasts** show a preferred orientation parallel to the strain–slip cleavage s_2. This suggests that these porphyroblasts grew while the s_2 cleavage was developing, or later. The few biotite flakes in Figure 8.2 do not follow the penetrative schistosity direction as closely as the chlorite and muscovite flakes, again suggesting that they grew later. Thus, from the textural features of the rocks in Figures 8.2 and 8.3 a history of interrelated metamorphic crystallisation and deformation for the low-grade part of the Furulund Schist can be inferred. In the earlier phase, the s_1 schistosity developed, and chlorite and muscovite were crystallising, while biotite crystallised during or after the formation of the strain–slip cleavage s_2. It would be incorrect to extrapolate this conclusion and suggest that biotite grew later than muscovite and chlorite throughout the Sulitjelma region. The regional picture can only be built up by studying the minor structures in the field, and the textures of thin sections from all parts of the area.

What of the variation in metamorphic grade? It has already been stated that local variations in grain size make this only a rough guide to metamorphic grade. The mineral assemblages of the schists are more reliable. The rock shown in Figure 8.2 has the assemblage quartz + plagioclase + muscovite + chlorite + opaque minerals + biotite while that in Figure 8.3 has the assemblage quartz + muscovite + biotite + epidote + opaque minerals. Compare these with the assemblage of the rock illustrated in Figure 8.4, which has the assemblage biotite + muscovite + plagioclase + quartz + garnet + opaque minerals although it has approximately the same composition as the rock shown in Figure 8.3.

The change in the mineral assemblages, the incoming of garnet, is a

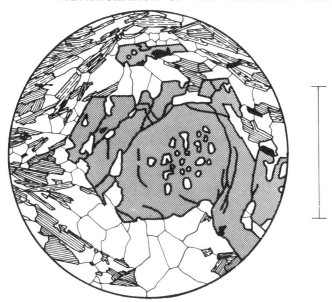

Figure 8.4   Garnet-mica schist. Scale bar 1 mm.

consequence of increasing metamorphic grade. Garnet is not found at all in
the southeastern part of the area. In the north and west it is not found in all
specimens of the Furulund Schist, but only in bands relatively rich in both
aluminium and iron. The garnets are clearly visible in hand-specimen
because they grow as porphyroblasts and, even when the proportion of
garnet is low and the crystals of comparatively small size (about 0·1 mm),
they form small knots which may be seen by the naked eye to disturb the
usually uniform schistosity. The rock shown in Figure 8.4 is a comparatively coarse-grained garnet–mica schist, the garnet having a diameter
of about 1·5 mm; the change in direction of the penetrative schistosity
round the garnet porphyroblast is particularly marked.

The first appearance of garnet in the more aluminium- and iron-rich
bands of the Furulund Schist can be quite accurately located on the ground,
because of the fairly continuous rock exposure in this recently glaciated
arctic region. In Figure 8.1, in the centre of the map, two bands of schist are
shown intercalated among the Sulitjelma Amphibolites. When the southern
strip of the two is followed on the ground from east to west across the small
lake Otervatn, the schist is found to be free of garnets on the eastern side,
while on the western side quite common small porphyroblasts can be seen.
Thus, for the comparatively uniform rock composition of this schist strip,
almandine garnet is present in the mineral assemblage on the western side
of the lake, but not on the eastern. The conditions of metamorphism have
changed, i.e. the metamorphic grade increased, by a critical amount, so

that garnet formed on one side but not on the other. A line, known as an **isograd** or line of equal metamorphic grade, crosses the strip of schist at this point (Fig. 8.1). When the progressive metamorphic sequence in the Sulitjelma Amphibolites is discussed, it will be shown that the incoming of garnet is not the only significant change in the metamorphic assemblages occurring at this line.

Isograds are so important in the discussion of progressive regional metamorphism that it is worth pausing in the description of the Sulitjelma area to define them further. The definition of an isograd was first given by Tilley (1924b). As has already been explained, an isograd may be defined by the *first* appearance of a new mineral with increasing metamorphic grade in a progressive regional metamorphic sequence. It need not be seen in all horizons of any one stratigraphical unit, but only in those of a suitable composition. For example, it has already been shown that almandine garnet only appears in the more iron- and aluminium-rich layers of the Furulund Schist group of Sulitjelma. However, it is only possible to map a line of outcrop of an isograd, like that shown in Figure 8.1, if horizons suitable in composition for garnet to appear at the isograd are fairly common.

The isograd line recorded on a map, like the boundaries between formations marked on conventional geological maps, is only a two-dimensional representation of the isograd, which is a surface when viewed in three dimensions. The shape of the isograd surfaces in an orogenic belt is of great interest from a geotectonic point of view; but unfortunately in many areas it is difficult to locate the positions of the isograds on a map with precision because of poor exposure or scarcity of rock horizons of suitable composition, and so the shape of the isograds may be impossible to determine.

The AFM triangles of Figure 8.5a and b represent the changes in the mineral assemblages on either side of the garnet isograd. The close similarity in chemical composition between garnet and chlorite suggests that garnet has formed by the breakdown of chlorite, and the triangles show this. However, the metamorphic reaction cannot be this simple, because the Fe/Mg ratios of almandine and chlorite are different, and because at Sulitjelma chlorite persists above the almandine isograd, although its proportion in the schists falls rapidly. If the AFM triangles of Figure 8.5 are right, the isograd is marked by a metamorphic reaction involving a change in the Fe^{+2}/Mg ratios of chlorite and biotite. This would be a **sliding reaction** as defined in Chapter 13.

The relationship between isograds and metamorphic reactions makes isograds of particular significance in metamorphic petrogenic studies. Reactions such as the one above may be investigated by experiments in mineral synthesis in the laboratory, and also to some extent by theoretical physical–chemical calculation, leading to a determination of the conditions

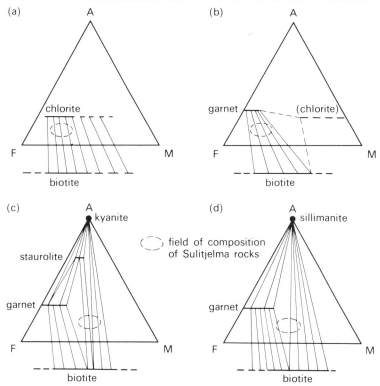

Figure 8.5 AFM triangles showing the progressive change in the mineral assemblages of pelitic rocks at Sulitjelma with increasing grade. (a) Biotite zone. (b) Garnet zone. (c) Kyanite zone (early assemblage of Figure 8.7). (d) Sillimanite zone (late assemblage of Figure 8.7).

of temperature, pressure and activity of various chemical components at the isograd surface during the period of metamorphism. This will be discussed at more length in Chapter 13.

In the author's view, however, it is wrong to *define* isograds by the metamorphic reactions which they represent, although enthusiasts for this approach to metamorphic petrogenesis, such as Winkler (1976), try to do so. This ignores the fact that it is difficult to determine the chemical equation for a metamorphic isograd, even though the isograd may be easily mapped in the field. In many cases different equations of apparently equal plausibility are proposed by different workers. As more information is acquired not only about the mineral assemblages of metamorphic rocks but also about the rock and mineral compositions, it may well be possible to write more widely acceptable equations; but for the present it seems better to stick to a definition of isograds in terms of field recognition of the first appearance of minerals, in rocks of a particular chemical composition.

Strictly speaking, an isograd need not only be defined by the incoming of a new metamorphic mineral with increasing grade. It may be defined by the disappearance with increasing grade of a mineral which is stable at low grade but not at high grade. For example, chlorite is absent from the highest-grade schists of Sulitjelma, so theoretically an isograd might be mapped by the disappearance of chlorite. However, because chlorite does not form porphyroblasts like garnet, and because it frequently appears as a secondary alteration product formed after the main metamorphic event, such an isograd is impracticable to map. It is also possible to define isograds by the incoming or outgoing of *pairs* of minerals with increasing grade, but again this has not proved practicable at Sulitjelma.

Most isograds which have been mapped in regional metamorphic rocks are recognisable by hand-specimen study of the rocks concerned, like the garnet isograd at Sulitjelma. In some cases isograds have been mapped by comparison under the petrological microscope or by X-ray diffraction techniques of the metamorphic assemblages from hundreds of carefully located samples from a metamorphic terrain. The student will appreciate what a time-consuming task this is.

The garnet schist of Figure 8.4 also reveals some interesting aspects of the growth of garnet porphyroblasts and the textural evolution of the higher-grade parts of the Furulund Schist. The garnet crystal is **poikiloblastic**, the inclusions being of quartz. There is a distinct zonal arrangement of the quartz inclusions, and in Chapter 15 it will be shown that this is matched by compositional zoning of the garnet, although this is not detectable under the petrological microscope. In the centre of the garnet crystal there are small inclusions, then a zone free of inclusions and near the outside there are larger inclusions, some of them connected with the granoblastic matrix of quartz and plagioclase outside the garnet crystal. The outer inclusions show how the garnet has developed a poikiloblastic texture by growing faster along the boundaries between quartz grains than it grows through the interior of the grains. Although the garnet is poikiloblastic, its grain boundaries with the quartz and feldspar crystals show a very strong tendency to run parallel to the edges of a hexagon, which represents a cross section through the rhombic dodecahedron (form {110} of the cubic holosymmetric crystal class) which is a common crystal habit of members of the garnet group of minerals. This indicates that, during growth, the surface energy of the growing garnet crystal was relatively large. Contrast this mode of development of poikiloblastic texture with that of cordierite (Fig. 5.3).

Even when the partial replacement of the enclosed quartz grains by garnet is taken into account, it is clear from the spacings of the *centres* of the inclusions that the grains enclosed in the core of the porphyroblast were part of a granoblastic fabric which was finer-grained than the present fabric. There is also an indication by the elongation of the quartz grains that

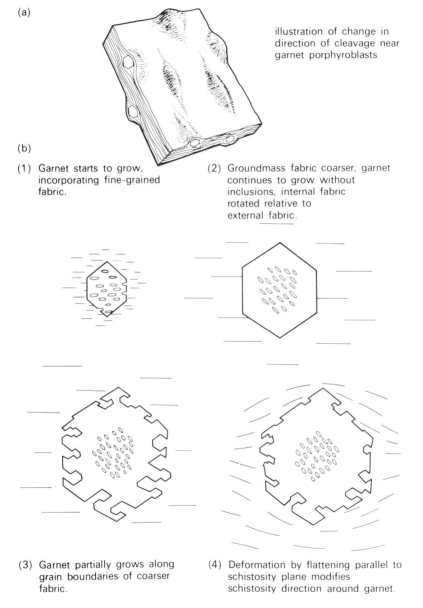

(a)

illustration of change in direction of cleavage near garnet porphyroblasts

(b)

(1) Garnet starts to grow, incorporating fine-grained fabric.

(2) Groundmass fabric coarser, garnet continues to grow without inclusions, internal fabric rotated relative to external fabric.

(3) Garnet partially grows along grain boundaries of coarser fabric.

(4) Deformation by flattening parallel to schistosity plane modifies schistosity direction around garnet.

Figure 8.6 (a) Hand-specimen textural relationships of garnet porphyroblasts in Furulund Schist. (b) Model for the evolution of garnet porphyroblasts.

this early fabric showed a foliation. The plane of this foliation (s_i) now makes a large angle with the plane of schistosity of the specimen (s_e), indicating that during the growth of the garnet porphyroblast there was a relative rotation of the fabric inside the garnet (s_i) compared with the fabric outside (s_e). The finer grain size of the internal fabric may suggest that it formed when the metamorphic grade was lower than the grade at the time when the external fabric crystallised; but the unreliability of grain size as an indicator of metamorphic grade has already been mentioned.

The schistosity direction in the micas tends to swing round the garnet porphyroblast in an eye-like pattern. The three-dimensional form of this disruption of the schistosity is shown in Figure 8.6a. Note that the spindle-shaped 'eyes' round the garnets define a lineation, as well as the foliation represented by the schistosity. This disruption of the uniform direction of the schistosity is not due to pushing aside of the pre-existing foliation by the growing garnet as the older textbooks of metamorphic petrology (e.g. Harker 1932) suggest. It has already been shown that the garnet por-phyroblast grew by replacing the existing fabric rather than pushing it aside. Figure 8.6b shows a mechanism for the formation of the texture by the intensification of the foliation of the schist during the last phase of growth of the garnet, the intensification arising by flattening deformation parallel to the plane of the schistosity accompanied by appreciable exten-sion along the direction of lineation. Independent evidence of a late stage of extensional deformation in the Furulund Schist is provided by some of the boudinage structures seen.

HIGHER METAMORPHIC GRADES IN SCHIST

Although the Furulund Schist is pelitic in composition, it is relatively rich in CaO, which is commonly present in the lower-grade schists in the form of calcite. At higher grades, the calcite tends to have reacted with phyl-losilicates, so that the aluminium-rich mineral is plagioclase feldspar. The characteristic mineral assemblages of the staurolite, kyanite and sillimanite zones of Barrow are seen in the more aluminium-rich Lappheleren Schist which lies above the Sulitjelma Amphibolites (Fig. 8.1). Figure 8.7 shows a thin section from a sample of this schist formation. It is from among the highest-grade regional metamorphic rocks found in the Sulitjelma area and the hand-specimen shows its comparatively coarse grain size and re-cognisable porphyroblasts of almandine garnet, staurolite and kyanite. The schist has an incipient crenulation cleavage, as well as its schistosity, and the thin section has been cut at right angles to both crenulation cleavage and schistosity. Notice that the magnification of the drawing in Figure 8.7 is less than that of Figures 8.2, 8.3 and 8.4.

The porphyroblast minerals kyanite and staurolite have not previously been described. Staurolite is seen in hand-specimen as stumpy prismatic

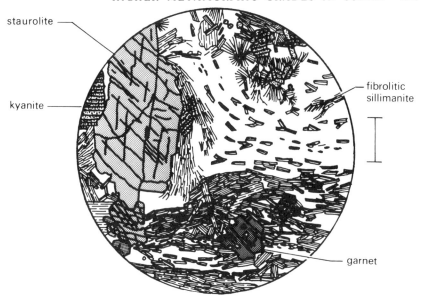

staurolite

kyanite

fibrolitic
sillimanite

garnet

Figure 8.7 Kyanite-staurolite-mica schist. Plagioclase-quartz groundmass not
subdivided into individual grains. Scale bar 1 mm.

crystals of a dark brown colour, not developing the cross-shaped twins by
which it is frequently recognised. The kyanite occurs as typical bladed
crystals up to 5 cm long, the planes of the blades lying in the schistosity
direction and the elongation being parallel to the crenulation lineation. The
blades show the characteristic pale blue colour of kyanite. Under the
microscope a third new mineral is encountered, the prismatic aluminium
silicate sillimanite. In this specimen it appears as both the needle-like
textural variety known as **fibrolite** and the more characteristic prismatic
form (Fig. 8.8).

Both kyanite and sillimanite have the same chemical composition,
Al_2SiO_5. This is the same as the composition of andalusite which has
already been described from contact metamorphic aureoles (Ch. 5). Now
that all three polymorphs of Al_2SiO_5 have been encountered, it is appropriate to discuss the optical properties which distinguish them under the
petrological microscope and also the petrogenetic significance of these
minerals.

Although all three minerals are orthosilicates with independent SiO_4
tetrahedra, their optical and physical properties are strongly influenced by
chains of aluminium ions surrounded by oxygen ions in six-fold octahedral
coordination. Therefore all three polymorphs show a tendency to prismatic
crystal habits. Andalusite and sillimanite are orthorhombic, kyanite tri-
clinic. The optical properties which they all share, and which distinguish

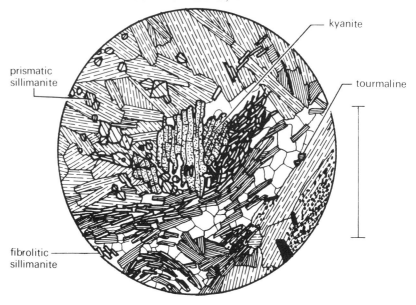

Figure 8.8 Kyanite-sillimanite textural relationships in the same rock as Figure 8.7. Scale bar 1 mm.

them from the micas, pyroxenes and amphiboles with which they may be associated, are comparatively high positive relief and low birefringence. Sillimanite may show second order red or blue interference colours, but otherwise the colours are first order grey or yellow. The lack of anomalous interference colours due to dispersion distinguishes the aluminium silicates from minerals of the epidote group. The student should look out for aluminium silicates in all coarse-grained metamorphic rocks of pelitic composition, especially if garnet or staurolite are present and plagioclase feldspar is not abundant. The easiest way to distinguish the three aluminium silicate polymorphs under the microscope is to search for sections at right angles to the prism direction. The shape and pattern of cleavages of these basal sections are shown in Figure 8.9. The basal section of sillimanite is especially distinctive, and fibrolite fibres like those shown in Figure 8.8 may show it when examined under the highest magnification. The optic signs and optic axial angles are also useful distinguishing properties and good interference figures may be obtained even on small grains.

The petrogenetic significance of these minerals is due to their identical chemical composition, and to the fact that petrographic studies suggest that all three polymorphs have stability fields lying within the range of conditions of regional and contact metamorphism. This should be clear already, because andalusite has been encountered as a stable mineral in the higher-grade parts of contact aureoles, and now kyanite and sillimanite have been seen in a high-grade regional metamorphic setting. If the mutual stability relationships of the three polymorphs could be determined, they might be a very useful indicator of pressure and temperature conditions in metamorphism. For this reason a good deal of effort has been put into synthetic mineral studies to determine the stability field of each polymorph in the pure Al_2SiO_5 system (Ch. 13).

Petrographic study leads to uncertainty whether the Al_2SiO_5 minerals in high-grade schists always grew within their stability fields of temperature and pressure. Two of the polymorphs are often found together, as in Figures 8.7 and 8.8. Several instances have been recorded where all three polymorphs are found in one rock. It is not likely that the temperature and pressure conditions at the climax of metamorphism fell exactly on a **univariant** line between two stability fields, as shown in Figure 13.3, and still less likely that they fell exactly at the **triple point** where all three lines join. It is more likely that temperature–pressure conditions changed so that one or more lines were crossed, and Al_2SiO_5 minerals survived **metastably** outside their own stability fields.

In the case of the rock shown in Figures 8.7 and 8.8, the textural relationships suggest that kyanite grew earlier, followed by first prismatic

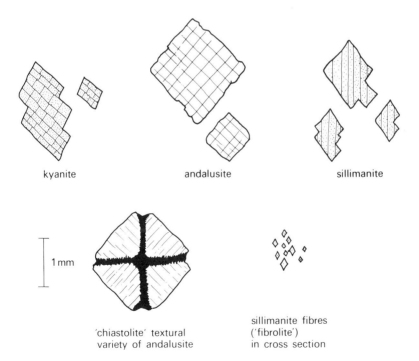

Figure 8.9   Basal sections of Al_2SiO_5 polymorphs.

sillimanite and then fibrolitic sillimanite (see Fig. 8.8). The kyanite appears to be somewhat corroded and the sillimanite prisms show idioblastic outlines, as do the fibres when examined under high magnification. The sillimanite is not a straightforward pseudomorphous replacement of the kyanite crystals. Textural relationships between aluminium silicate minerals are usually somewhat ambiguous in this way.

The rock shown in Figure 8.7 therefore contains the minerals quartz, biotite, muscovite, kyanite, staurolite, sillimanite, almandine garnet and one or more opaque minerals. This is not an equilibrium assemblage list however, because either kyanite or sillimanite must have been out of equilibrium when the main fabric of the rock crystallised. As the kyanite appears to have crystallised before the sillimanite, it may be that it has survived from an earlier, slightly lower-grade period in the metamorphic history. Like the kyanite, the staurolite shows rather corroded crystals, so it may be that two mineral assemblages are represented in the rock, the earlier one being quartz + biotite? + staurolite + kyanite + muscovite? + opaques (the minerals with question marks are inferred, because all the minerals of that kind recorded in the thin section probably belong to the later fabric), and the later one quartz + biotite + muscovite + garnet + sillimanite + opaques. These mineral assemblages and an estimated rock composition field are shown in the AFM triangles of Figure 8.5c and d.

Although because of the field relations between the aluminium-rich Lappheleren Schist and the calcium-rich Furulund Schist it is not possible to map isograds for the incoming of staurolite, kyanite and sillimanite at Sulitjelma, there are many other progressive regional metamorphic sequences in which this has been done. The porphyroblastic character of staurolite and kyanite often make mapping of these isograds possible by study of hand-specimens in the field.

PROGRESSIVE REGIONAL METAMORPHISM OF THE SULITJELMA AMPHIBOLITE GROUP

The Sulitjelma Amphibolites are a sequence of metamorphosed lavas, intrusive sheets and pyroclastic rocks, predominantly of basic igneous composition. They overlie the Furulund Schist structurally, and are cut across by the isograds of progressive regional metamorphism.

It is not possible to recognise isograds of regional metamorphism in the amphibolites in the field. There is the same tendency for grain size to increase with increasing metamorphic grade, but the local variability of grain size is even more marked in the Sulitjelma Amphibolites than in the Furulund Schist. One noticeable feature is that in the fine-grained bands of the amphibolites the predominant colour is light green in the low-grade parts of the area and dark green in the high-grade parts. The reasons for this change will be discussed later.

As the name **amphibolite** implies, the Sulitjelma Amphibolites contain essential calcic amphibole and plagioclase feldspar and, in coarse-grained specimens, both may be distinguished in hand specimen. In the field, primary igneous features are frequently recognisable. Layers are frequently found with pyroclastic fragments, flattened and elongated by tectonic deformation but still recognisable. Pillow structures also occur. The massive layers show chilled margins at their contacts, and prismatic jointing normal to the contacts, indicating that they are a large number of separate sheet-like intrusions. Many of the more massive amphibolites display relict porphyritic texture, with recrystallised **pseudomorphs** after plagioclase feldspar. In thin section the amphibolites show calcic amphibole, plagioclase feldspar, minerals of the epidote group and, in much smaller amounts, quartz, chlorite, garnet, apatite, sphene and opaque minerals. Many contain calcite or dolomite.

A number of progressive changes occur with increasing metamorphic grade in the mineral assemblages of the Sulitjelma Amphibolites. These will be illustrated by considering two rocks, one from a comparatively low-grade part of the area, the other from a high-grade part.

Figure 8.10 shows a thin section of a massive pale green amphibolite belonging to the lower part of the Sulitjelma Amphibolite group of the Swedish part of Sulitjelma. It comes from an area in which pillow structure has been recognised but not from an outcrop showing pillow structure. The

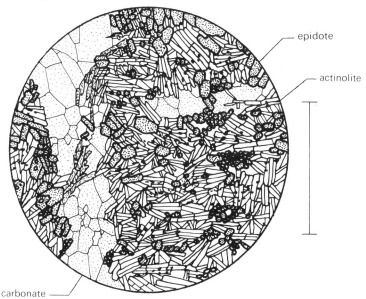

epidote

actinolite

carbonate

Figure 8.10   Thin section of massive portion of pillow lava, Swedish Sulitjelma. Specimen location is east of the area shown in Figure 8.1, in the Sulitjelma Amphibolite. Scale bar 1 mm.

rock has the mineral assemblage epidote + actinolite + carbonate + plagioclase + quartz and is cut by veins about 1 mm across of carbonate, clinozoisite and chlorite, although chlorite does not appear to be an essential mineral of the rock itself.

The amphibole is colourless and occurs as clusters of radiating prismatic crystals. It is not easy to distinguish actinolite from hornblende under the microscope, but in this case the prismatic habit of the amphibole makes determination of the maximum extinction angle easy, and it is 20°, which is a value more characteristic of actinolite than hornblende. The amphibole is very pale green, not sufficiently pleochroic for its pleochroic scheme to be determined. There is no preferred orientation of the actinolite prisms, which interlock to give the rock its fine-grained and tough character. Between the actinolite prisms are interstitial crystals of plagioclase. They may be identified because one or two crystals show wedge-shaped lamellar twins due to strain. The extinction angle is low, below 15°, and the grains have refractive indices close to that of the mounting medium ($1 \cdot 53$). This suggests that the plagioclase is of albite composition (An_{0-10}). Conoscopic study supports this conclusion by showing that the cleaved grains are biaxial positive. There are also a few grains with no cleavage and slightly higher R.I.s and birefringence, which give uniaxial positive interference figures in convergent light and are thus shown to be quartz.

The other major mineral of this rock is epidote. It occurs as stumpy prismatic crystals with rounded outlines and a large range of grain sizes. It has appreciably higher positive relief than the amphibole and under crossed polars displays the anomalous interference colours which are characteristic of the epidote group of minerals. The colours go up to second order green, indicating that the epidote group mineral in the main part of the rock is iron-bearing epidote. It is possible to compare this with the low-iron clinozoisite in the veins with calcite, which shows only low first order interference colours. These anomalous colours are caused by dispersion, in both epidote and clinozoisite, and are particularly striking in the second order green and in the low first order colours which are prussian blue or pale brown, as well as grey. The rock contains accessory sphene which has even higher R.I.s than the epidote, high birefringence, and a brown colour which is visible in the larger grains.

This rock illustrates well the differences in petrographic study needed for metamorphic rocks in comparison with igneous ones. The petrological significance of the type of amphibole present means that its extinction angle must be determined. The scarcity of twinning in plagioclase means that more attention must be paid to properties such as refractive index, cleavage and optic sign, in an effort both to demonstrate the presence of plagioclase and to determine its composition. It is usually necessary to study metamorphic rocks using the highest power of magnification of the microscope.

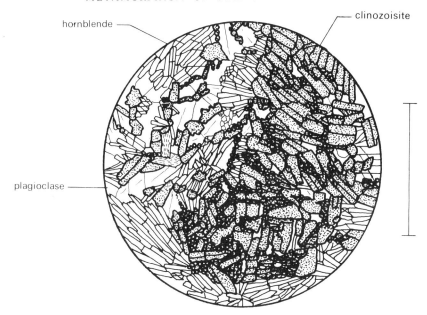

Figure 8.11   Amphibolite from massive portion of porphyritic lava-flow, including relict phenocryst. Scale bar 1 mm.

The rock shown in Figure 8.11 comes from a higher-grade part of the Sulitjelma Amphibolites. In hand-specimen it is darker green than the previous rock, probably because the amphibole is hornblende, not actinolite. Its most conspicuous feature is the presence of white patches which are pseudomorphs after plagioclase phenocrysts of the original basalt. The specimen comes from a massive sheet in the upper part of the Sulitjelma Amphibolite group. Such massive sheets with relict porphyritic texture are characteristic of the Sulitjelma Amphibolite group over much of the Swedish part of Sulitjelma.

In thin section **pseudomorphs** after calcic plagioclase are equally conspicuous and consist of crystals of clinozoisite and plagioclase. The clinozoisite shows low first order birefringence colours, anomalous blue and brown due to dispersion, and has a prismatic habit and a wide range of grain sizes. The surrounding part of the amphibolite shows the assemblage hornblende + plagioclase + epidote + sphene + opaques. If the rock is studied under low magnification, it can be seen that the hornblende crystals occur in parallel or radiating clusters, with patches of epidote and plagioclase between them. These probably represent respectively the pyroxene and calcic plagioclase of the groundmass of the original igneous rock, which would therefore have been a medium-grained basalt or dolerite with plagioclase phenocrysts.

The maximum extinction angle (27°) suggests that the amphibole in the groundmass is hornblende not actinolite. The epidote mineral is identified as epidote by its anomalous interference colours, second order bire-fringence and negative optic sign. The presence of two different members of the epidote group in the phenocrysts and groundmass indicates that during metamorphism aluminium and ferric iron were not free to reach equal concentrations throughout the volume of the rock. The higher amount of ferric iron available in the groundmass resulted in the crystallisation of relatively ferric iron-rich epidote, while the lack of either ferric or ferrous iron in the phenocrysts, which were originally labradorite or andesine plagioclase, means that iron-free clinozoisite formed. Therefore, during metamorphism there must have been a chemical potential gradient of Fe^{3+} ions at the edges of the recrystallising phenocrysts assuming the two epidote group minerals grew at the same time. Some evidence for this is apparent in the thin section. The clinozoisite crystals near the edges have a slightly higher birefringence than those in the core, implying that there was a limited amount of diffusion of Fe^{3+} ions into the phenocrysts.

The plagioclase of both phenocrysts and groundmass shows quite abun-dant lamellar twinning and the composition may be determined by the maximum symmetrical extinction method in both cases, and in the groundmass also by the Carlsbad–albite combined extinction angle method. These measurements show that the composition of the plagioclase is uniform, in both phenocrysts and groundmass, at An_{28}.

When the mineral assemblages of the rock of Figure 8.10 and that of the groundmass of Figure 8.11 are compared, there are two major differences which are due to the difference of metamorphic grade, not the small difference in chemical composition. One is that hornblende has taken the place of actinolite as the essential calcic amphibole, the other that calcic oligoclase has replaced albite as the essential plagioclase.

Systematic study of the amphibolites of the Sulitjelma area by Henley (1970) has shown that the changes from actinolite to hornblende, and from albite to oligoclase plagioclase, take place abruptly as the almandine garnet isograd of the Furulund Schists is crossed. The changes at the isograd are not recognisable in the amphibolites in the field. Henley has also demon-strated that in amphibole and plagioclase-bearing bands of the Furulund Schist, there is a similar abrupt change in the composition of the amphibole and plagioclase at the garnet isograd. Therefore, the isograd is not only a garnet isograd, it is also an oligoclase and a hornblende isograd. Isograds for oligoclase and hornblende have been mapped in other progressive regional metamorphic sequences, but do not necessarily coincide with the garnet isograd as at Sulitjelma.

The change from albite to oligoclase is particularly interesting. Henley's study shows that albite only occurs with a range of composition from An_0 to An_8 and that oligoclase shows a minimum anorthite content of An_{17}. The

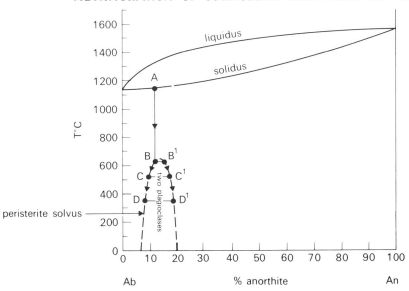

Figure 8.12   The system $NaAlSi_3O_8 - CaAl_2Si_2O_8$, illustrating the peristerite solvus. The liquidus and solidus curves are for a pure, dry system under a pressure of 100 kPa. The temperatures of crystallisation of plagioclase feldspars in igneous rocks are lower, but the curves are of the same form. A plagioclase which crystallised with composition A would cool to temperature B as a homogeneous crystal. At temperature B it would separate into two plagioclase phases with compositions B and B'. As the temperature fell further, the gap in composition between the two coexisting plagioclases would widen, e.g. to C and C' at 530° C and D and D' at 350° C.

range of plagioclase compositions from An_8 to An_{17} is absent. Although at the high temperatures at which plagioclase crystallises from igneous magmas there is complete solid solution between albite and anorthite, this is not the case at the considerably lower temperatures of recrystallisation of amphibolites at Sulitjelma. Albite and oligoclase are immiscible phases. X-ray studies on igneous plagioclases within the composition range An_8 to An_{17} also shows that they are sub-microscopic intergrowths of albite and oligoclase, and these plagioclase intergrowths have been given the name **peristerite**. Thus there is an immiscibility region in the albite–anorthite phase diagram, below the solidus temperature, which is known as the 'peristerite solvus' (Figure 8.12). The jump in plagioclase compositions as the isograd is crossed gives an indication of the temperature of metamorphism: the wider the gap, the lower the temperature of metamorphism (Cooper 1972).

This does not explain the general tendency for the anorthite content of plagioclase in amphibolites to increase with increasing metamorphic grade.

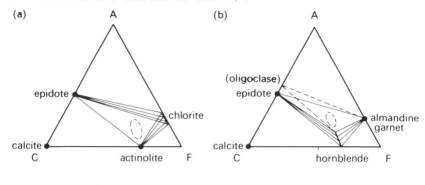

 field of composition of Sulitjelma Amphibolites

Figure 8.13 ACF triangles to illustrate progressive metamorphism of basic igneous rocks at Sulitjelma. (a) Assemblages below garnet-oligoclase isograd. (b) Assemblages above garnet-oligoclase isograd. Oligoclase should not strictly be shown on ACF triangles, but is indicated by dotted tie-lines in (b). From Vogt (1927).

This has been widely observed and is a useful indication of metamorphic grade. The replacement of albitic by anorthitic plagioclase suggests that calcium and aluminium are being introduced to the plagioclase. The replacement of actinolite by hornblende implies that aluminium and sodium are being supplied to the amphibole. The most likely source of calcium and aluminium is the breakdown of minerals of the epidote group with increasing grade; and the sodium in the hornblende might be supplied by partial breakdown of the albite component in the plagioclase. Because of this, the anorthite content of plagioclase can only be used as a guide to metamorphic grade in rocks which also contain epidote minerals or other Ca-bearing species. As the metamorphic grade increases, epidote breakdown apparently takes place gradually, and the anorthite content of the plagioclase continues to increase above the minimum permitted by the peristerite solvus gap until, in the highest-grade amphibolites of Sulitjelma, the anorthite content of the plagioclase is about An_{50} and epidote group minerals are absent. There also appears to be a progressive, continuous change in the composition of the hornblende reflected in a stronger green colour with increasing grade.

The mineralogical changes occurring in the amphibolites with increasing metamorphic grade are summarised in the ACF triangular diagrams of Figure 8.13. The incoming of garnet in the Furulund Schist is also illustrated on these triangles.

The lack of tectonic fabrics in the amphibolites means that they do not give textural evidence of crystallisation of the metamorphic minerals in relation to the deformation history, unlike the schists. But they make up for this in textural interest because they frequently show textural features related to their primary igneous origin.

METAMORPHIC GRADE

Now that several examples of progressive metamorphic sequences in contact and regional metamorphic rocks have been described in terms of complete mineral assemblages (except for the opaque minerals), it is worth trying to define metamorphic grade more precisely than in Chapter 2. In pelitic rocks, the reactions which have been demonstrated to occur with increasing metamorphic grade are dehydration reactions: chlorite has been progressively replaced by garnet in the Furulund Schist, and chloritoid and chlorite by andalusite, cordierite and biotite in the Skiddaw Slates. In these cases, increase in metamorphic grade is marked by progressive dehydration of the mineral assemblages. In the siliceous dolomites of the Beinn an Dubhaich Aureole, by·contrast, the increase of metamorphic grade is marked by progressive decarbonation of the mineral assemblages. Thus in metamorphosed sedimentary rocks, metamorphic grade is defined by the progressive loss of primary volatile components. Sedimentary diagenesis also involves the loss of volatile matter and metamorphism can be seen as an extension of diagenesis by further volatile loss (Ch. 10).

The *relative* character of metamorphic grade was emphasised in Chapter 2. In accordance with that chapter, the metamorphic grade of a sedimentary rock may be defined as the relative amount of primary volatile components which has been lost from the mineral assemblage. The mineral assemblage of a low-grade rock has lost a smaller proportion of the primary volatile components, the mineral assemblage of a high-grade rock a greater proportion. H_2O and CO_2 are not the only volatile components which may be lost. Coal loses CH_4 and other volatiles with increasing 'rank' (i.e. sedimentary diagenesis) and also when it undergoes metamorphism (Miyashiro 1973, pp. 232–33).

In metamorphosed igneous rocks, the definition of metamorphic grade is more difficult. The high-grade members of the Sulitjelma Amphibolite sequence have less H_2O than the low-grade members, but the original rock was presumably virtually anhydrous. The Sulitjelma Gabbro nearby (Fig. 8.1) is still mostly anhydrous, but its field relationships indicate that it probably remained at the same temperature and pressure as the surrounding schists and amphibolites for a very long time. It has remained effectively unmetamorphosed because H_2O did not penetrate far into the large intrusion. Therefore to undergo metamorphism at all (at least under the temperature and pressure conditions of metamorphism at Sulitjelma), igneous rocks need to have volatile material introduced. In many geological settings (Ch. 10, Ch. 14) this happens rather readily. In such cases metamorphic grade may be defined by stating that metamorphic rocks of low grade contain more introduced volatile material in the metamorphic mineral assemblage than do high-grade rocks.

Metamorphism does occur in rocks with no volatile components, for example the Fra Mauro breccias of the Moon (Ch. 6). In these cases the temperature of onset of metamorphism is much higher than the highest temperature reached in the Sulitjelma sequences. In the Fra Mauro breccias metamorphic grade is simply defined by the temperature of recrystallisation of the metamorphic assemblages (Taylor 1975). The rigorous definition of metamorphic grade in rocks which are volatile-free throughout metamorphism lies beyond the scope of this book.

Exercise

The following analyses are average values from specimens of Furulund Schist, of higher metamorphic grade than the garnet isograd (analyses 1 to 3) and from the Lappheleren Schist (analysis 4). These analyses, like those at the end of Chapter 7, were performed by X-ray fluorescence analysis, but in this case the totals have not been altered to add up to 100%. Because H_2O has not been determined, and all the iron is quoted as FeO, the totals are not significant and are not given. Plot the analyses onto an ACF diagram, making the assumptions that all K_2O is in muscovite and all Na_2O is in plagioclase. Treat all the iron as FeO, as it is quoted in the analyses. These assumptions ignore the presence of biotite in the mineral assemblages, but it is probably not present in a large enough quantity to make nonsense of the diagram.

	(1)	(2)	(3)	(4)
SiO_2	62·0	62·5	64·4	61·4
TiO_2	1·07	0·99	0·78	1·12
Al_2O_3	16·7	16·2	12·8	18·2
FeO*	8·0	7·2	5·3	8·5
MnO	0·24	0·14	0·09	0·11
MgO	5·1	4·6	3·9	3·3
CaO	2·3	3·6	7·4	0·5
Na_2O	1·7	1·7	1·9	1·0
K_2O	2·9	3·0	1·8	4·3

*Total iron given as 'FeO .

Mineral assemblages
1. Quartz + plagioclase + muscovite + biotite + garnet + opaques.
2. Quartz + plagioclase + muscovite + biotite + garnet + hornblende + opaques.
3. Quartz + plagioclase + muscovite + biotite + hornblende + opaques.
4. Quartz + plagioclase + muscovite + biotite + staurolite + opaques.

(Unpublished analyses by K. J. Henley, gratefully acknowledged; each an average of four analyses of different specimens.)

Using the compositions of minerals in Figure 8.13 (take staurolite as A 69%, F 31%), sketch the tie-lines to indicate the mineral assemblages. Note that the results are broadly correct in spite of the sweeping assumptions made at the beginning. Estimate the maximum aluminium content of the hornblende and compare it with your results from the Exercise at the end of Chapter 5. The purpose of this Exercise is to show that even with incomplete analytical data it may be possible to plot triangular diagrams and obtain useful results. Notice also that ACF diagrams may be used for pelitic rocks when these contain Ca-bearing minerals such as hornblende, clinopyroxene or epidote.

9

Regional metamorphic rocks of Palaeozoic orogenic belts, II

The sequence of mineral changes seen with increasing metamorphic grade in the Furulund Schist Group of Sulitjelma is typical of progressive regional metamorphism of many pelitic rocks. It is not the only type, however. An example of a different type of regional metamorphic sequence is found in schists of the Dalradian Supergroup in the Connemara district of County Galway, western Ireland. The geology of this area has been studied by

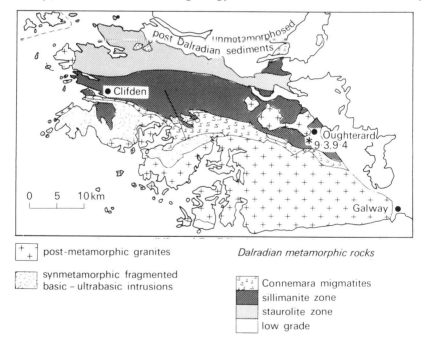

Figure 9.1 Geological sketch map of Connemara, western Ireland. Mainly based upon Leake (1970b), with metamorphic zone boundaries based on several sources (see text). Locations of specimens in Figures 9.2, 9.3 and 9.4 shown.

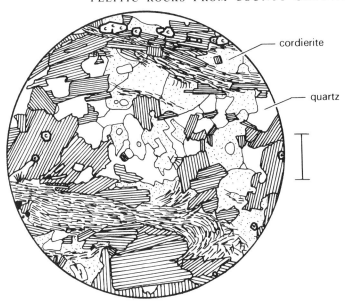

cordierite

quartz

Figure 9.2   Cordierite-sillimanite schist, Oughterard, Co. Galway. Scale bar 1 mm.

several workers under the leadership of Leake (Leake 1958, 1970a, Badley 1976, Yardley 1976). Figure 9.1 is an outline geological sketch map of the area, based on this work. The Dalradian rocks are interbedded quartzites, schists, marbles and basic volcanic rocks. Although the sequence is broken up by thrust faults which formed before metamorphism, marker horizons permit a reasonable correlation to be made with southwestern Scotland, indicating that the rocks were deposited at the very end of Precambrian time.

The Dalradian Supergroup is overlain unconformably by Lower Palaeozoic rocks whose age ranges down to the Tremadocian stage of the Ordovician (Skevington 1971). This unconformity is marked by a strong contrast in metamorphic grade. The Ordovician and Silurian rocks are low-grade slates, whereas the Dalradian includes high-grade schists and gneisses. High-grade metamorphism of the Dalradian clearly predated the uplift, erosion and deposition which caused the unconformity (Dewey *et al*. 1970).

PELITIC ROCKS FROM COUNTY GALWAY

Figure 9.2 shows a cordierite–sillimanite schist from the Dalradian of Galway. This rock has the mineral assemblage: cordierite + quartz + sillimanite + biotite + plagioclase + opaques + apatite. The sillimanite

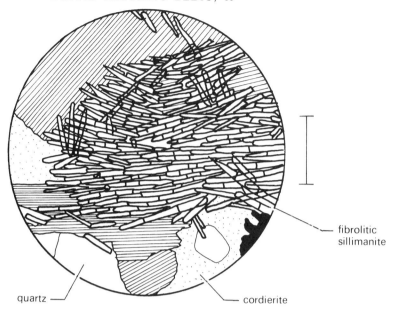

Figure 9.3 Enlarged drawing of part of rock shown in Figure 9.2. Scale bar 0·1 mm.

occurs as mats of sub-parallel **fibrolite** needles, parallel to the cleavage planes of biotite. Figure 9.3 shows an enlargement of the end of a bundle of needles of sillimanite, showing how the needles grow at the grain boundaries of the coarser-grained biotite and cordierite. This indicates that the growth of the sillimanite continued after biotite and cordierite had ceased to crystallise. Examination of the sillimanite needles in this thin section under the higher magnifications of the microscope shows a variation in grain size. Surprisingly, the needles on the outside of the mats are coarser-grained than those on the inside. It is tempting to think that this indicates that sillimanite grew while the metamorphic grade was increasing, but the association of the sillimanite mats with biotite flakes suggests an alternative explanation. The sillimanite probably grows over the biotite because a similarity in crystal structure encourages small sillimanite crystals to begin growing on biotite grain boundaries (Chinner 1966). This initiation of the growth of new crystals is called **nucleation**. Because conditions for nucleation on the surface of biotite are so favourable, many small crystals of sillimanite nucleate, but as growth goes on, only some of them continue to grow. Fibrolitic sillimanite often shows textures which indicate the importance of nucleation processes in its development. A mineral which nucleates easily may grow **metastably** outside its stability range. If this has occurred, of course the mineral should not be included in an assemblage list.

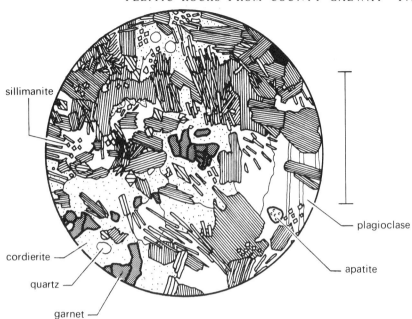

Figure 9.4 Garnet-cordierite-sillimanite schist, Oughterard, Co. Galway. Scale bar 1 mm.

In this case, the abundance of sillimanite, and the indication from the variation in grain size of the needles that it continued to grow for quite a long time in the textural evolution of the rock, are taken to indicate that it was a member of the equilibrium mineral assemblage. The reader will appreciate the subjective nature of this judgement. Such subjective judgements often have to be made in the study of all three aluminium silicate minerals, which nonetheless have great importance in the interpretation of the metamorphic history of many pelitic rocks. The mineral assemblage of this rock resembles that of the high-grade hornfelses of the Skiddaw Aureole described in Chapter 5, except that the aluminium silicate is sillimanite, not andalusite. This resemblance to a contact metamorphic assemblage will be discussed further in the next section of the chapter.

Figure 9.4 shows a schist in which garnet porphyroblasts have partially broken down. One example is at the centre of the drawing, another at the lower left. The garnet has broken down to cordierite, sillimanite, biotite and quartz, and it is significant that nowhere in the thin section is garnet in contact with biotite. Sillimanite clearly coexists with biotite, overgrowing it and also being included within it. The garnet is thus a **relict**, surviving from an earlier stage of metamorphism of the rock. Regional descriptions of pelitic rocks from Connemara mention other relict minerals, such as stauro-

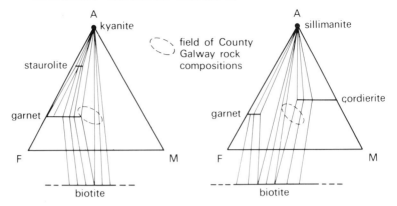

Figure 9.5 AFM triangular diagrams to show the change between early and late mineral assemblages of pelitic rocks, Co. Galway.

lite, kyanite and andalusite. The triangular diagrams in Figure 9.5 show the stable mineral assemblage of the rock in Figure 9.4, and also a possible earlier set of mineral assemblages from which the garnet might have survived. The published accounts of relict minerals (e.g. Yardley 1976) suggest that staurolite and kyanite are likely to be members of the relict mineral assemblages in this part of Connemara. The two diagrams clearly suggest a change from metamorphic assemblages characteristic of the Barrow zones towards mineral assemblages more like those of Skiddaw or Comrie.

VARIATION IN THE CONDITIONS OF METAMORPHISM IN SPACE AND TIME

Throughout the Dalradian rocks of County Galway there is a tendency for metamorphic grade to increase from north to south, becoming highest in a zone of **migmatites**, known as the Connemara Migmatites, in the south of the area (Fig. 9.1, Leake 1970b). The Connemara Migmatites are frequently in contact with a very large fragmented sheet of basic and ultrabasic rocks, which has no contact aureole. Figure 9.1 also shows the variation in metamorphic grade in County Galway based upon a few areas for which isograds have been published (e.g. Yardley 1976) and others where it is possible to guess at the approximate positions from published petrographic descriptions. The isograds appear to be defined by mineral assemblages which came into equilibrium during the second of the two stages in the metamorphism described above, but it is difficult to be certain of this throughout the area because of the limited amount of published information about the metamorphic rocks.

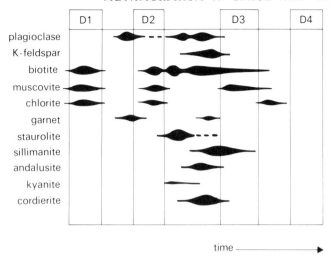

D1, D2, D3, D4 are deformation episodes

Figure 9.6   Metamorphic mineral growth in relation to four episodes of deformation. From Yardley (1976).

While the distribution of metamorphic conditions in space through the Dalradian rocks of County Galway is sparsely documented, the variation in time in certain areas has been very carefully described. By comparing the development of axial planar cleavages related to different episodes of folding with the growth of metamorphic minerals, it has been possible to work out a sequence of metamorphic mineral growth in relation to episodes of deformation. Figure 9.6 shows a sequence obtained by Yardley (1976) from the central part of the area. D1, D2 and D3 are episodes of deformation which can be recognised throughout the area shown in Figure 9.1. Staurolite, kyanite and garnet grew during D2 or between D2 and D3, while andalusite, sillimanite and cordierite grew later between D2 and D3 or during D3. If the order of growth of metamorphic minerals is comparable throughout Connemara, making due allowance for variation in grade, it appears there was an early phase of metamorphism under conditions comparable to those at Sulitjelma, and a later phase of metamorphism under different conditions.

The later phase of metamorphism produced metamorphic assemblages similar to those seen in Dalradian pelitic rocks of Banffshire, Scotland, in the Buchan district of the county. It is therefore called a **Buchan-type** metamorphic sequence, in contrast to the earlier phase which is called a **Barrow-type** metamorphic sequence after Barrow, who first described it.

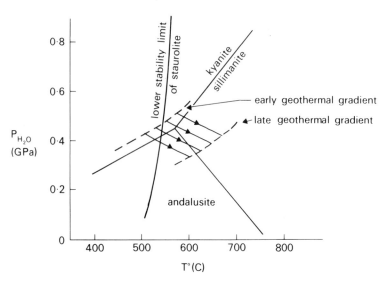

(small arrows show possible temperature – pressure paths of individual rocks)

Figure 9.7   Early and late geothermal gradients, Co. Galway. From Yardley (1976).

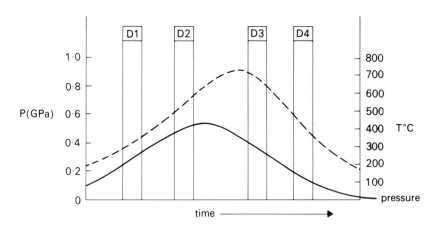

Figure 9.8   Variation of temperature and pressure with time. Sillimanite zone, Co. Galway.

What is the reason for the difference between the two progressive regional metamorphic sequences in County Galway? Figure 9.7 shows the stability ranges of certain minerals (Ch. 13) and plots possible temperature–pressure gradients during metamorphism in the early and late metamorphic phases, respectively. During the early Barrow type metamorphism, the rate of increase of temperature with depth in the Earth's crust was lower than the corresponding rate during the later Buchan-type metamorphism. The **geothermal gradient** during Barrow metamorphism was lower than during Buchan metamorphism. Figure 9.8 shows the variation of temperature and pressure with time in the sillimanite zone rocks of County Galway (like those in Figures 9.2, 9.3 and 9.4) assuming that this interpretation is correct.

The resemblance of the metamorphic assemblages of the later (Buchan) phase of metamorphism to contact metamorphic assemblages has been mentioned already. Leake (1970a) has made the ingenious suggestion that it really is a very large-scale type of contact metamorphism. He pointed out that the Connemara Migmatites could very reasonably be interpreted as the products of partial melting of the Dalradian sediments (Fig. 7.8) brought about by the great amount of heat released during the crystallisation and cooling of the great basic–ultrabasic intrusive sheet mentioned earlier. Dispersal of the same heat through the Dalradian sediments could be responsible for the pattern of isograds shown in Figure 9.1, which was established during the Buchan metamorphism. The immense scale of the heating by the basic–ultrabasic sheet would explain the absence of a *local* contact aureole, like that of the Sulitjelma Gabbro (Ch. 2).

This raises a problem of the category of such metamorphism (Ch. 1). If Leake is right, it might be argued that it should be called contact metamorphism, as it is in the paragraph above. However, the metamorphism has occurred over an area of approximately 230 km^2, and is detectable at least 15 km from the contacts of the basic–ultrabasic sheet. Because the metamorphism thus covers a region, rather than being local, it is regarded by the present author as regional metamorphism even though Leake's suggestion may well be correct.

Exercise

The following analyses are of pelitic rocks from one small area in the central part of Connemara (Fig. 9.1). Plot them on an AFM diagram and sketch in tie-lines to indicate the mineral assemblages.

	(1)	(2)	(3)	(4)	(5)
SiO_2	49·73	48·00	41·88	46·38	52·53
TiO_2	0·39	0·46	1·38	1·03	1·02
Al_2O_3	27·50	27·83	29·20	30·37	26·76
Fe_2O_3	1·66	1·40	1·15	1·98	1·45
FeO	6·89	9·36	10·76	9·74	7·86
MnO	0·08	0·23	0·16	0·15	0·18
MgO	1·61	1·95	3·09	2·09	2·31
CaO	0·53	0·90	0·98	0·59	0·61
Na_2O	0·46	0·62	1·56	1·42	1·25
K_2O	6·26	5·52	4·63	2·40	2·71
P_2O_5	0·18	0·09	0·23	0·16	0·17
H_2O+	4·20	3·19	4·45	3·13	2·69
S	0·10	0·40	0·21	0·18	0·04
Total	99·59	99·95	99·68	99·62	99·58

Mineral assemblages

1. Muscovite + sillimanite + biotite + quartz + garnet.
2. Muscovite + staurolite + garnet + quartz + biotite.
3. Muscovite + biotite + staurolite + plagioclase + garnet.
4. Staurolite + muscovite + quartz + biotite.
5. Staurolite + quartz + muscovite + biotite + garnet.

(Analyses from Leake (1958), reproduced with permission.)

One rock is the 'odd man out'. Which is it, and why?

Note Allow for the sulphur by removing $\frac{1}{2}[S]$ from the $[FeO]$, in calculating F. This assumes that the sulphur is present in pyrite, FeS_2.

(Analysis 1 is the 'odd man out', containing garnet + sillimanite in place of staurolite. Thus it comes from above the staurolite isograd, while the other rocks come from below. It does come from further south in the area than the other three rocks.)

10

Regional metamorphic rocks of Cainozoic orogenic belts

The examples of progressive regional metamorphic sequences which will be described in this chapter all come from the Alpine orogenic belt of southern and central Europe. This is an extremely complex orogenic belt and includes progressive metamorphic sequences of several different types. Because of its comparatively recent formation, erosion has not cut so deeply into the deformed rock sequences as in the Caledonian orogenic belt so that in some areas a transition from unmetamorphosed sediments into the lowest grades of regional metamorphism can be studied. The strong topographical relief enables the shape of isograd surfaces to be determined in three dimensions in some cases. Finally, some progressive metamorphic sequences are of a type not yet described in this book. The examples to be described will continue to concentrate upon progressive metamorphic sequences in pelitic rocks and in basic igneous rocks.

A regional account of the structure of the Alps is beyond the scope of this book. The reader with no knowledge of the subject is referred to an excellent short introduction by Umbgrove (1950) and to the summary by Read & Watson (1975b). Figure 10.1 is a small-scale map of the western and central Alps, showing some of the zones of progressive regional metamorphism which formed in Cainozoic times, the principal tectonic divisions, and the areas of rocks metamorphosed in Palaeozoic or Precambrian times.

PROGRESSIVE REGIONAL METAMORPHISM OF PELITIC ROCKS ABOVE THE AAR MASSIF

In the northern part of the central Alps in eastern Switzerland, Mesozoic and Cainozoic sedimentary rocks lie above a large mass of older rocks, the Aar Massif. Some of the sediments lie unconformably upon Palaeozoic metamorphic rocks and intrusive igneous rocks. These crystalline rocks

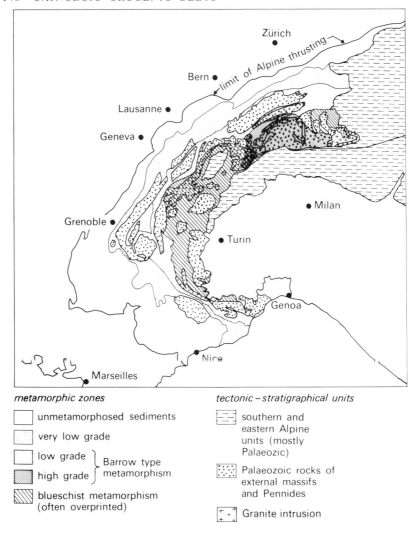

Figure 10.1   Sketch map showing Alpine metamorphism in the western and central Alps. Much simplified from Frey *et al.* (1974).

and the overlying sedimentary sequence together are referred to as **autochthon**. The sedimentary rocks above the autochthon have been tectonically
transported on thrust planes from south to north. This process has already
been discussed in Chapter 6. The transported rocks lie in several **nappes**
which are grouped into a major tectonic unit, the Helvetic Nappes. The
rocks of both the autochthon and the Helvetic Nappes range in age from

Permian to Eocene. Two rock compositions predominate: limestones and shales.

The lowest parts of the complex pile of sediments overlying the Aar
Massif are low-grade regional metamorphic rocks. The metamorphism has
not changed the mineral assemblages of the carbonate rocks. The pelitic
sediments, by contrast, have undergone considerable mineral and textural
changes to become slates. In several places they have been quarried for
roofing material. The upper parts of the Helvetic Nappes show no regional
metamorphism and therefore the area is an admirable one for studying the
onset of regional metamorphism in pelitic rocks (Frey 1974). In these rocks,
the phyllosilicate minerals recrystallised and changed their composition
during regional metamorphism. The changes are most appropriately
studied in the laboratory by X-ray diffraction techniques, but microscopic
study of thin sections enables the minerals concerned to be recognised and
shows something of the complicated textural development in these rocks.

Figure 10.2 shows a thin section of a specimen of roofing slate from
autochthonous sediments of Eocene age on the northern side of the Aar
Massif. The original sediment was a sandy mud and the bedding is picked
out by layers with more quartz grains. These grains retain their clastic
shapes. The white mica flakes are very small and are indistinguishable from

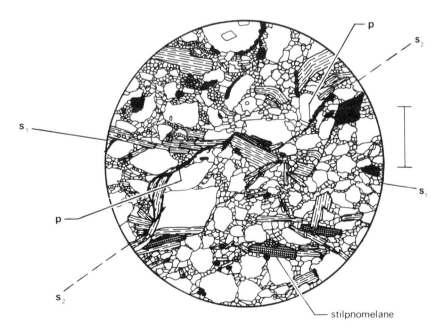

Figure 10.2   Thin section of roofing slate from Sernftal, Canton Glarus, Switzerland. Scale bar 0·1 mm.

muscovite under the petrological microscope. Because X-ray studies show that they are not muscovite, the informal name **sericite** will be used to describe them. The mineral assemblage of this rock is quartz + sericite + calcite + stilpnomelane + opaques. It is a marly (lime-rich) slate. Stilpnomelane is a mineral not previously described in this book. It is a brittle mica with a complex chemical formula, notably rich in iron, and is characteristic of low-grade rocks between unmetamorphosed sediments and low-to-medium grades of **Barrow-type** regional metamorphism. In the pelitic rocks of the central Alps, the appearance of stilpnomelane defines an isograd, which is shown on Figure 10.1. Under the petrological microscope, stilpnomelane may be difficult to distinguish from biotite. It is pleochroic from brown to yellow and the flakes have parallel extinction with the slow vibration direction parallel to the cleavage. In the rock shown in Figure 10.2 the stilpnomelane can be identified because it has a slightly less perfect cleavage than biotite, and does not show the characteristic mottling close to the extinction position between crossed polars. In progressive regional metamorphic sequences in pelitic rocks, stilpnomelane does not coexist with biotite but appears at a much lower metamorphic grade.

There are two distinct grain sizes of sericite in this rock. The larger grains define a penetrative slaty cleavage, which cross-cuts the bedding although this cannot be seen in Figure 10.2. The smaller grains have a less strong preferred orientation. There is a later, strongly defined non-penetrative cleavage in the rock, which is parallel to the axial surfaces of minor folds. In the outcrop, these folds can be seen to fold both the bedding and the earlier slaty cleavage. Some clastic quartz grains show **pressure solution** on the non-penetrative late cleavage surfaces (points **p** in Figure 10.2). Although in general the quartz grains have not recrystallised, it is apparent from the shapes of grains adjacent to the **non-penetrative** cleavage surface that parts of some grains have been dissolved away against the surface. The local higher pressure on the grain against the developing cleavage surface is thought to have promoted the solution of the quartz in intergranular fluid. This mechanism is important in the deformation of sediments and very low-grade metamorphic rocks, but probably becomes less significant when most of the grains in the rock are undergoing metamorphic recrystallisation during deformation. In this particular rock, the non-penetrative cleavage appears to have developed after the maximum metamorphic grade was reached, during the development of the penetrative cleavage. Pressure solution of the quartz grains does not appear to have happened at maximum metamorphic grade.

Figures 10.3 and 10.4 show a rock of higher metamorphic grade from the autochthon on the southern side of the Aar Massif. The mineral assemblage is quartz + plagioclase (albite) + potash feldspar + sericite + chlorite + graphite. This list is based upon X-ray determinations of the minerals by C.

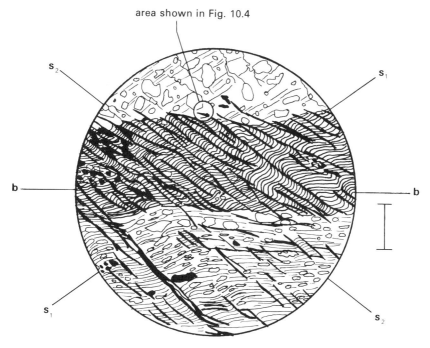

Figure 10.3 Thin section of metamorphosed interbanded graphitic phyllite and sandy slate, Lötschental, Valais, Switzerland. Scale bar 1 mm.

Taylor (personal communication). The plagioclase is not recognisable under the microscope because it is untwinned. The rock is different in composition from that of Figure 10.2 in that the sand component is arkosic and the fine-grained groundmass does not contain carbonate. The textural development of this rock also shows two distinct cleavages, the earlier one penetrative, the later non-penetrative. Pressure solution of clastic quartz grains is visible on the later non-penetrative cleavage surfaces. The opaque graphite has a recognisable platy habit and tends to be parallel to the preferred orientation direction of the sericite grains. Figure 10.4 shows part of the rock under the same magnification as Figure 10.2. This illustration shows that the non-penetrative cleavage surfaces (s_2) run along the axial planes of microfolds in the sericite. The concentration of opaque graphite in the s_2 surfaces is probably a pressure solution effect. It has been left in the cleavage surfaces after quite a lot of sericite has been removed by pressure solution because it is extremely insoluble. This process is analogous to the formation of stylolites in limestones. The mineral assemblage of this rock is comparable with those of the lowest-grade pelitic rocks found in the Sulitjelma region of Norway (Ch. 8, Fig. 8.2).

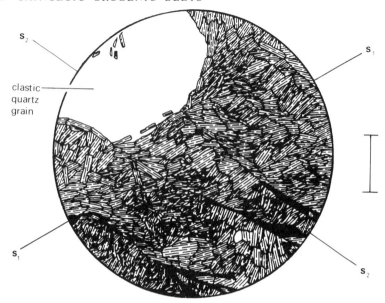

Figure 10.4   Enlarged part of Figure 10.3 (indicated). Scale bar 0·1 mm.

It was stated earlier that the sericite in these low-grade Alpine metamorphic rocks is not muscovite. It is a mixed-layer phyllosilicate with some layers in the crystal structure of muscovite while others are of the clay mineral illite. As the metamorphic grade increases, the ratio of muscovite to illite increases, and the sericite loses bound H_2O from its crystal structure. In the rock of Figure 10.3 the sericite has reached a composition which is regarded as falling within the muscovite range, although it is a variety relatively rich in magnesium which is given the name **phengite**. The progressive change in the composition and structure of the sericite has been studied by X-ray diffraction techniques. As the proportion of illite in the sericite diminishes, the spacing of the planes of silicon and aluminium ions in the crystals becomes more regular. This causes the X-ray reflections from prepared mounts of sericite powder to become more sharply defined. The sharpness is expressed as a number known as the **illite crystallinity** (Dunoyer de Segonzac 1970). In sericites from unmetamorphosed shales the illite crystallinity is greater than 7·5, in sericites of the stilpnomelane zone it is 4·0–7·5, and in the chlorite zone of higher grade than the stilpnomelane zone it is less than 4·0. The illite crystallinity of the rock in Figure 10.3 is 3·2. That of the specimen in Figure 10.2 has not been determined experimentally, but comparison with specimens from the nearby Linth valley suggest that the value should lie in the range 5·5–6·0 (Frey & Hunziker 1973).

SSE   NNW

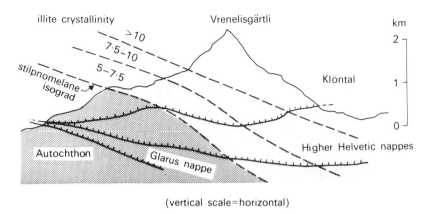

(vertical scale=horizontal)

Figure 10.5 Isograds shown on a cross section along the western side of the Linth valley, Canton Glarus, Switzerland. From Frey and Hunziker (1973).

Frey and Hunziker studied specimens from an area with vertical relief of more than 3000 m and were therefore able to determine the shape of **isograd** surfaces. The isograds dip north approximately parallel to the upper surface of the Aar Massif. Figure 10.5 shows the form of the isograd surfaces in cross section along the west side of the Linth valley. The mineral assemblages and the illite crystallinity values correlate well with one another as indicators of metamorphic grade. Several specimens have also been dated radiometrically by the potassium–argon method applied to glauconite, stilpnomelane and biotite (Ch. 14). The dates are similar at all metamorphic grades. It is often thought that isograds have the arched structure shown in Figure 10.5, but it has seldom been so convincingly demonstrated. The isograds cut across the thrust planes between the nappes, indicating that most of the nappe transport occurred before metamorphism. This conclusion directly contradicts that of Schmid (1975) mentioned in Chapter 6. Further research will be needed to resolve the problem.

BLUESCHISTS FROM THE WESTERN AND CENTRAL ALPS

In the inner parts of the Alpine orogenic belt are rocks of Mesozoic age which are clearly metamorphic, unlike the rocks of the Helvetic Nappes which until recently were usually described as unmetamorphosed sediments. The tectonic units to which these rocks belong are collectively known in Switzerland as the Pennide Nappes. The Pennide Nappe Mesozoic successions differ from those of the Helvetic Nappes in including basic igneous rocks, both extrusive basaltic lavas and intrusive gabbros.

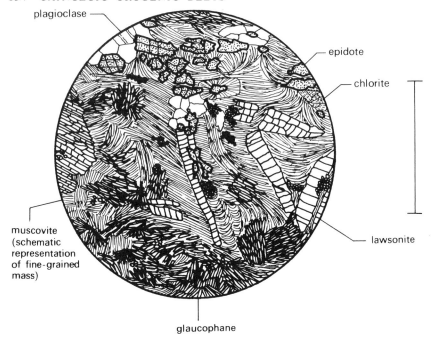

plagioclase

epidote

chlorite

muscovite
(schematic
representation
of fine-grained
mass)

lawsonite

glaucophane

Figure 10.6 Metamorphosed basalt, from Pic Marcel, near Guillestre, France. Harker thin section collection Cambridge, no. 118474. Scale bar 1 mm.

The metamorphic sequences of the Pennide Nappes are varied. Rocks of pelitic composition are quite abundant and in some places show progressive metamorphic sequences of typical Barrow type (Fig. 10.1). There is also a different type of sequence, which is seen best in basic igneous rocks. These are metamorphosed to medium- or even coarse-grained schistose rocks. These could be regarded as amphibolites, because they usually contain amphibole, plagioclase feldspar and very often epidote. The amphibole however is remarkable in being blue in colour, and so these rocks are called **blueschists**.

Figure 10.6 shows a thin section of a typical blueschist from the equivalent tectonic unit to the Pennide Nappes, but in France. The specimen comes from rocks whose field relationships and structure show that they are submarine basalt flows. Pillow structure is preserved. The minerals in this rock, in order of abundance, are muscovite, glaucophane, albite, epidote, lawsonite, quartz and chlorite. This includes a number of minerals which have not yet been introduced, and will not be familiar to the reader from igneous rocks. Glaucophane is a sodic amphibole $Na_2Mg_3Al_2Si_8O_{22}(OH)_2$. Most natural glaucophanes are solid solutions between glaucophane with the formula above and riebeckite

$Na_2Fe_3^{+2}Fe_2^{+3}Si_8O_{22}(OH)_2$. The intermediate members of the series are called crossite. The distinctive feature of glaucophane, crossite and riebeckite is strong pleochroism. In glaucophanes and glaucophane-rich crossites the pleochroic scheme is α colourless, β lavender blue, γ blue. The lavender blue colour when light vibrates in the β direction is diagnostic of glaucophane and crossite (there are varieties of hornblende which show blue colour in certain vibration directions). It is the glaucophane or crossite amphibole crystals which give the rock its blue colour in hand-specimen. This may not be obvious (the author has failed to recognise blueschists in hand-specimen) but in thin section they cannot be mistaken.

Lawsonite is a lime- and aluminium-bearing hydrated silicate, $CaAl_2(OH)_2(Si_2O_7)H_2O$. This formula is equivalent to that of anorthite plagioclase $CaAl_2Si_2O_8$ with the addition of $2H_2O$, and lawsonite takes the place of the anorthite component of the plagioclase of the primary basalt. It is orthorhombic and optically positive, which distinguishes it from epidote, with which it is easily confused, especially as the two minerals often occur in the same rock, as here.

A feature of this rock is that the metamorphic minerals show a considerable variety of textures. For example, glaucophane is present both as large crystals (one appears on the left) and as radiating clusters of needles. Similarly, there are two distinct sizes of epidote crystals. This textural variety makes it uncertain that the rock is in a state of equilibrium. The mineral list may therefore not be a metamorphic assemblage. It does seem likely that glaucophane and lawsonite grew together, because they are found in contact with one another and because the two kinds of glaucophane crystal suggests that glaucophane was growing through quite a long time in the textural evolution of the rock, probably including the time when lawsonite grew. The importance of the coexistence of glaucophane and lawsonite will be explained shortly.

Textural evidence for disequilibrium is a frequent feature of blueschists. It is often associated with confusing field relations, in which patches of medium-to-coarse-grained blueschists may be found among pelitic rocks apparently showing rather uniform low-grade metamorphism. Progressive metamorphic sequences in which isograds may be mapped in the field are seldom found, although in some areas isograds have been identified by the study of numerous thin sections. However, some specimens of blueschists do seem to show equilibrium mineral assemblages and, in many others, it is possible to infer that some minerals coexisted in equilibrium during part of the time of metamorphism, as with the lawsonite and glaucophane in Figure 10.6.

Figure 10.7 shows a coarser-grained blueschist from the Pennide Nappes of the Zermatt region of Switzerland. It contains glaucophane and epidote, but not lawsonite. Unlike the rock in Figure 10.6, however, this rock contains metamorphic pyroxene. This pyroxene is a sodium-rich variety of

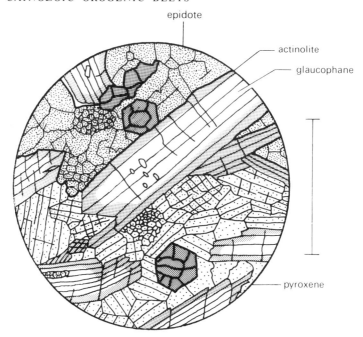

epidote

actinolite

glaucophane

pyroxene

Figure 10.7 Glaucophane-epidote schist, Riffelhaus, near Zermatt. Harker thin section collection, Cambridge, no. 19435. Scale bar 1 mm.

augite known as **omphacite**, whose optical properties resemble those of augite except that omphacite has a pale green colour in plane-polarised light. The rock also contains subhedral to euhedral garnets. Although these are indistinguishable optically from the garnets of the schists of Sulitjelma, chemical analysis shows them to be richer in the magnesium garnet component pyrope ($Mg_3Al_2Si_3O_{12}$). The rock is much coarser-grained than that in Figure 10.6, and might almost be considered to represent an equilibrium mineral assemblage. The amphiboles in the rock show that this is not the case. Individual prismatic amphibole crystals are obviously zoned compositionally, their cores showing the blue and lavender colours of glaucophane, their rims the dark green of actinolite. The birefringence of the rims is also visibly greater.

It is not easy to work out an equilibrium mineral assemblage for this rock. It appears most likely that there might have been an early assemblage garnet + omphacite + glaucophane + epidote, which subsequently locally broke down to albite + actinolite (there is a small rim of albite in the upper left of the drawing). If this is the case, the early assemblage is perhaps a transitional one between blueschist and **eclogite**, a pyrope-rich garnet + omphacite rock (Ch. 12). The later assemblage represents a period of

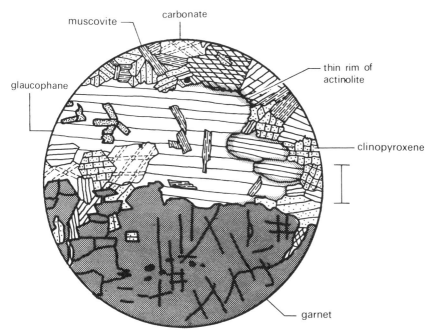

Figure 10.8 Glaucophanitic eclogite, Pfulwe, Zermatt. Harker thin section collection, Cambridge, no. 92882. Scale bar 1 mm.

retrograde metamorphism, i.e. later metamorphism at a lower metamorphic grade.

Figure 10.8 shows a rock which exhibits not two but three phases of metamorphism. A large garnet crystal is in contact with omphacite. The omphacite appears to be undergoing replacement by a large glaucophane crystal associated with a carbonate mineral, probably calcite or dolomite. Small prisms of epidote and flakes of muscovite appear to be growing inside the glaucophane crystal, which also has a narrow rim of actinolite, like the crystals in Figure 10.7. This rock could be interpreted as having three successive mineral assemblages:

high grade   garnet + omphacite
↓
↓   glaucophane + calcite
↓
low grade   actinolite + muscovite + epidote

In this case each suggested mineral assemblage is of lower grade than the preceding one, suggesting a lengthy time sequence of retrograde metamorphism.

Because of the zoning of minerals, and in some cases the variation in habit and grain size of minerals (like Fig. 10.6), the recognition of assemblages of metamorphic minerals in blueschists is more difficult, and therefore more liable to error, than in the regional metamorphic rocks described in Chapters 8 and 9. One point needs to be mentioned. If a metamorphic rock shows two-stage metamorphism, with an early metamorphic assemblage breaking down to a later one, the second assemblage will depend not only on the rock composition and the conditions of metamorphism, but also on the extent and character of the breakdown of the first assemblage. For example, if the early assemblage is garnet + omphacite and the later one glaucophane + lawsonite ± epidote, the relative proportions of glaucophane and lawsonite and/or epidote will depend upon the compositions of the garnet and omphacite, and also the relative proportions in which they break down.

CONDITIONS OF FORMATION OF BLUESCHISTS

The disequilibrium textural features and the lack of field relationships indicating progressive metamorphic sequences in blueschists have been mentioned. It has been suggested that it is possible, in spite of these difficulties, to infer possible equilibrium mineral assemblages. Although the problem of estimating the conditions of metamorphism (temperature, pressure, fluid composition, etc.) from mineral assemblages will be treated in general terms in Chapter 13, the conditions for blueschists are so peculiar and interesting that they merit brief discussion here.

The blue amphiboles glaucophane and crossite, which are the obvious feature of blueschists, are of limited use for determining conditions of metamorphism. The reactions which might form them from actinolite during prograde metamorphism are complex, and rock composition influences the temperature and pressure at which they occur. Petrological experience shows that glaucophane-bearing rocks quite often occur which do not contain other characteristic minerals such as lawsonite or omphacite. The stability range of lawsonite is known; of interest to the discussion here is that it is a comparatively low temperature mineral, breaking down at temperatures ranging from 345 °C at 0·4 GPa to 430 °C at 1·0 GPa (Winkler 1976, p. 187). Low temperatures of metamorphism would also explain the widespread disequilibrium features of blueschists, because metamorphic reactions proceed more rapidly with increasing temperature and therefore it is more likely that evidence for uncompleted reactions would be seen in rocks metamorphosed at low temperatures.

The clearest evidence for pressures of metamorphism comes not from the blueschists themselves, but from rocks of different composition associated with them. In calcium carbonate marbles, it has been shown that the stable $CaCO_3$ mineral during metamorphism was not calcite, but aragonite.

In the best samples, this has a medium-to-coarse-grained granoblastic texture. In metamorphosed arkosic sandstones and quartz- and feldspar-rich greywackes, the sodic pyroxene jadeite ($NaAlSi_2O_6$) is stable in coexistence with quartz. The conditions under which the reactions

$$calcite \rightleftharpoons aragonite$$
$$albite + quartz \rightleftharpoons jadeite$$

have been well determined by laboratory experiment. The jadeite-forming reaction is discussed in Chapter 13. The crucial point for the present discussion is that for the temperature range indicated by the presence of lawsonite in blueschists, aragonite-bearing marbles and jadeite-bearing arkoses must have been metamorphosed at unusually high pressures (0·5–0·9 GPa for aragonite marble, 0·7–1·2 GPa for coexisting jadeite and quartz). To put these figures into perspective, the pressure at the base of continental crust is about 0·9 GPa (Holmes 1965, p. 921). There has been a great deal of argument about whether the assumptions behind these pressure estimates are valid, especially whether the minerals concerned occur in equilibrium assemblages. It has been shown that aragonite may form instead of calcite at much lower pressures if it is precipitated from fluids with other dissolved ions, rather than from solution in pure water. In spite of the arguments, the author of this book thinks that for some blueschists, including Alpine ones, the case for metamorphism at unusually low temperatures and high pressures is valid.

In rocks of basic igneous composition, it is believed that unusually high pressures during metamorphism can be inferred if glaucophane coexists with lawsonite, as in Figure 10.6. Basic igneous rocks with glaucophane and crossite but without lawsonite or omphacite are thought to be transitional types metamorphosed under pressures intermediate between those of Barrow type amphibolites and those of lawsonite-bearing blueschists. The term 'blueschist' is often confined to the high-pressure rock types.

This adds another example to the varying geothermal gradients during metamorphism which have been described earlier. If this view of their metamorphism is correct, blueschists formed under metamorphic conditions where temperature increased unusually slowly with increasing pressure. This conclusion is one of great importance for tectonic studies (Ch. 16).

Exercise

The rock whose AFM ratios were calculated in Table 4.1 and plotted in Figure 4.6a, is given at the head of the Table below, which also gives partial modal analysis of other pelitic rocks from the Lötschental area, Switzerland. The illite crystallinity values show that these specimens lie in a limited, low grade, range of metamorphic grade. Re-plot the analysis onto a triangular diagram showing A, K and N, as follows: $A = [Al_2O_3] - 0.33[FeO+MgO] - [Na_2O] - [K_2O]$, $K = [K_2O]$, $N = [Na_2O]$. Such a diagram is shown below, with approximate mineral composition fields indicated. From the data in the Table, show the range of composition of the Lötschental rocks over the diagram.

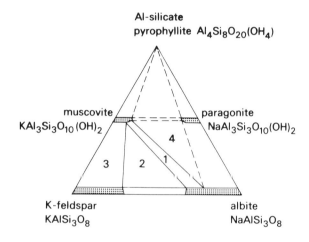

	I.C.	Musc.	Chlor. (percentages)	Pyrop.	K-f	Ab	Q	Cb	Op
Analysed Spec.	3·7	90	10	–	+	+	+	–	–
1	3·1	42	50	8	–	–	–	+ +	–
2	4·3	72	13	15	–	–	+ +	+	?
3	4·3	82	3	15	–	–	+	+	–
4	4·3	77	11	12	–	–	+	+	–
5	4·1	74	–	26	–	–	+	–	–
6	5·2	67	17	16	–	–	+	–	–
7	3·1	76	13	11	–	–	+	+ +	–
8	4·8	69	11	20	–	–	+	–	–
9	5·0	75	12	13	–	–	+	–	–
10	4·1	73	13	14	–	–	+	–	–
11	4·6	82	6	12	–	–	+	+	?
12	4·7	53	27	20	–	–	+	–	?
13	4·7	64	18	18	–	–	+	–	–
14	4·1	45	14	15	–	–	+	–	?
15	4·2	78	5	17	–	–	+	–	?
16	3·6	88	–	12	–	–	+	+ +	–
17	4·5	83	4	13	–	–	+	+	?
18	3·9	69	12	19	–	–	+	–	–
19	4·2	70	15	15	–	–	+ +	–	–
20	5·2	74	8	18	–	–	+ +	–	–
21	4·8	73	11	16	–	–	+ +	–	–
22	3·6	74	12	14	–	–	+ +	–	–
23	3·4	76	13	11	–	–	+ +	–	–
24	4·1	70	13	17	–	–	+ +	–	–
25	4·1	65	21	13	–	–	+	+ +	–
26	4·3	69	16	15	–	–	+	+ +	–
27	4·3	64	23	13	–	–	+	+ +	–

Key to table: I.C. – illite crystallinity, musc. – muscovite, chlor. – chlorite, pyrop. – pyrophyllite/paragonite, K-f – potash feldspar, Ab – albite, q – quartz, cb – carbonate, op – opaques. The first three columns give relative proportions of these three minerals; other columns: '–' – absent, + – present, + + – abundant, ? – possible. All results are based on X-ray diffraction. (Unpublished data collected by C. Taylor, gratefully acknowledged.)

11

Metamorphic rocks of the ocean floor

A large amount of information about the geology of the ocean floors has been collected in recent years. This has made it possible to generalise about the stratigraphy and petrology of the rocks in the shallow levels of the ocean crust at least, and to relate them to the processes of formation of oceanic crust by spreading from mid-oceanic ridges. In the 1960s metamorphic rocks were dredged from the deep ocean floor near large faults and the importance of metamorphism in the evolution of the ocean floor began to be appreciated (Miyashiro *et al.* 1971). Theories were advanced suggesting that the predominantly basaltic rocks of the ocean floor have a relatively rapid increase of metamorphic grade with depth because of the high rates of heat flow from the Earth's interior which are found near mid-oceanic ridges. Because the metamorphic rocks then available were dredge samples, it was not possible to check this idea directly; more recent improvements in techniques for drilling into oceanic crust, however, may well yield samples from beneath mid-oceanic ridges. An alternative way of checking the theories came when it was appreciated that there are masses of basic and ultrabasic rocks in orogenic belts which several independent lines of evidence suggest are fragments of oceanic crust.

The mechanisms of uplift and tectonic transport which have brought these fragments from the ocean floor onto continental crust are poorly understood; but the process has been given the name **obduction**. The masses of basic and ultrabasic rock are sometimes relatively little deformed internally, so the structure of the ancient ocean floor is preserved. In these cases metamorphic rocks from beneath the ocean floor may be studied by ordinary field and petrological techniques.

English-speaking geologists, following a very influential conference in the USA in 1972, have tended to use the name **ophiolite** suite for these fragments of ocean floor. This is unfortunate, as the name 'ophiolite' is used by European geologists for basic and ultrabasic rocks in the Alpine–Himalayan orogenic belt, irrespective of their origin on the ocean floor or elsewhere. This has led to a great deal of confusion, because the origin of many ophiolites, in the European sense, is obscure. The term 'ophiolite' will therefore not be used in this chapter. It is used in its descriptive sense, applied to Alpine rocks, in Chapter 16.

THE TROODOS BASIC–ULTRABASIC COMPLEX, CYPRUS

Probably the most famous of these basic–ultrabasic masses is that found in southwestern Cyprus, in the Troodos Mountains. Its location is shown in Figure 6.1. A brief account of the metamorphism of the rocks of the Troodos complex has been given by Gass and Smewing (1973).

The metamorphic rocks are mainly basic igneous rocks. In the shallower layers of the ancient oceanic crust they are submarine extrusive lavas with pillow structure, at deeper levels they are dolerite dykes, and at the deepest levels at which metamorphism occurs they are gabbros. Figure 11.1 shows a generalised column section through the ancient oceanic crust of Troodos, showing the primary igneous rock types and the superimposed progressive metamorphic sequence.

Figure 11.2 shows a metamorphosed basalt from the upper part of the Troodos sequence. The primary lath-shaped crystals of plagioclase, which crystallised from the igneous magma, remain unaltered. Between these crystals, there was originally a matrix of basaltic glass, but this has devitrified to radiating crystals of **smectite** intergrown in some parts of the rock with dusty grains of opaque minerals. 'Smectite' is a name for clay minerals of the montmorillonite group, which is used here in the informal way in

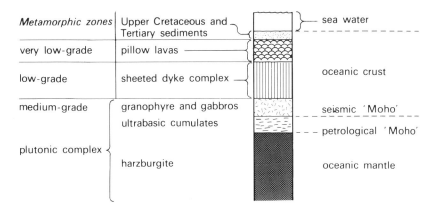

Figure 11.1 Generalised column section for the Troodos Complex, Cyprus, showing comparison with seismic layers of ocean floor (right) and progressive metamorphic sequence (left). From Gass and Smewing (1973).

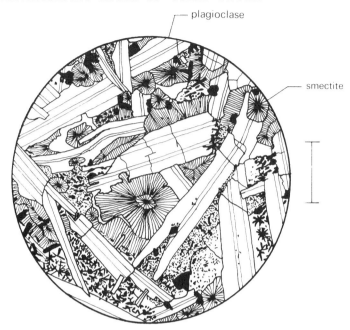

Figure 11.2 Basalt from very low-grade zone, Skouriotissa, Troodos. Scale bar 0·1 mm.

which 'sericite' was used in Chapter 10. The smectite grains are resolvable under the high magnifications of the petrological microscope. They are distinguishable from sericite micas by their pale yellow colour and lower refractive indices. The opaque mineral cannot be identified using transmitted light under the microscope, but has been determined by X-ray diffraction to be titanium-bearing magnetite. The sample comes from the massive part of a lava flow, which does not have amygdales containing the zeolite minerals which are especially characteristic of this part of the progressive metamorphic sequence. The presence of smectite replacing the basaltic glass matrix is also typical of the lowest-grade zone of metamorphism in the Troodos sequence. The texture clearly indicates that the rock is not in equilibrium and an assemblage list is not appropriate.

With increasing metamorphic grade, and depth beneath the ancient ocean floor, smectite is replaced by quartz and chlorite and the plagioclase feldspars become clouded owing to incipient breakdown. They still have a labradorite composition. Epidote and actinolite follow shortly after quartz and chlorite. Figure 11.3 shows a rock from this level. The mineral assemblage is plagioclase + actinolite + chlorite + quartz + opaques (titaniferous magnetite). This is similar to the low-grade metamorphosed basalt from Sulitjelma (Fig. 8.10) except that the plagioclase feldspar is anorthite-rich and an epidote mineral is not present, although epidote is

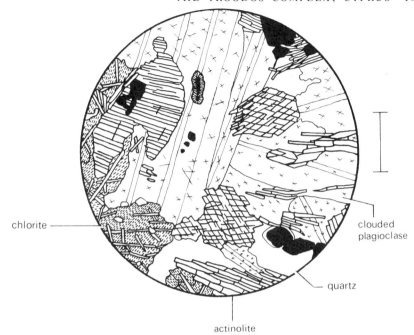

Figure 11.3 Basalt from low-grade zone, dyke in Sheeted Dyke complex, Phtery-koudhi. Scale bar 0·1 mm.

occasionally found in metamorphosed basalt from Troodos at this level. The plagioclase may retain its labradorite composition because it is not in equilibrium, having failed to break down to albite, epidote and quartz during metamorphism. In this case, the list of minerals given above is not an assemblage list. Alternatively, the plagioclase may be in equilibrium (Miyashiro 1973, p. 117) and the difference between the assemblages at Sulitjelma and Cyprus may be due to a higher geothermal gradient during metamorphism at Cyprus. The rock has a texture which suggests that it has an equilibrium mineral assemblage, unlike the rock in Figure 11.2 where the primary igneous plagioclase is obviously texturally distinct from the secondary metamorphic smectite.

With a further increase in metamorphic grade, actinolite is replaced by hornblende and chlorite disappears. Diopside-rich clinopyroxene joins the mineral assemblage. Figure 11.4 shows a rock of this grade. It has the assemblage plagioclase + quartz + hornblende + diopside + opaques. This rock comes from the lowest part of the sheeted dyke complex. Below this level the basic igneous rocks are increasingly replaced by ultrabasic rocks, which usually show little sign of metamorphism apart from local ser-pentinisation. Figure 11.5 shows an exception: a rock from a **xenolith** near the roof of a gabbro intrusion in the upper part of the Plutonic Complex. The coarser grain size compared with the rocks of Figures 11.2–11.4 is

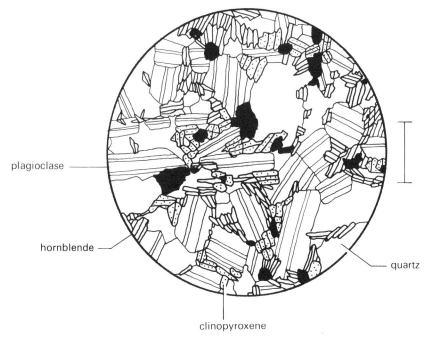

plagioclase

hornblende

quartz

clinopyroxene

Figure 11.4 Basalt metamorphosed in medium-grade zone, Sheeted Dyke complex, Lemithou. Scale bar 0·1 mm.

striking. The rock is a pyroxene hornfels and undoubtedly represents an equilibrium assemblage.

It may seem odd to include a contact metamorphic rock as the highest-grade member of a progressive regional metamorphic sequence. A rather similar approach was taken in Chapter 9, when it was suggested that the Connemara Migmatites and their associated Buchan-type progressive metamorphic sequence should be regarded as regional metamorphic rocks, rather than contact metamorphic rocks. As in the Irish example, it is suggested that the unusually high geothermal gradient associated with the metamorphism of the basic igneous rocks in the Troodos sequence might be due to the intrusion of a large volume of basic magma relatively high in the Earth's crust. The gabbro intrusion in which the xenolith of Figure 11.5 was found represents a small part of this intruded magma. Unlike the Irish example, where the basic intrusion appears to have been one large mass, the gabbroic layer in the Troodos sequence is made of a large number of smaller intrusions 1–5 km in diameter.

The assemblages of the higher-grade zones of the Troodos sequence are shown on ACF triangles in Figure 11.6. Because the compositional variation in the Troodos basic igneous rocks is small, the pattern of tie-lines can only be put on part of the ACF diagram.

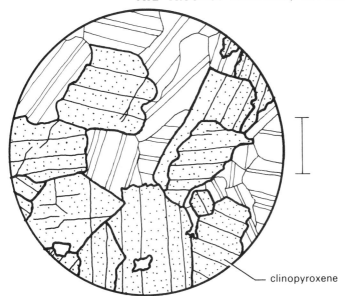

clinopyroxene

Figure 11.5 Pyroxene hornfels from xenolith of basalt in gabbro intrusion just below base of Sheeted Dyke complex, Ayios Demetrios. Scale bar 0·1 mm.

The rocks of the Upper and Lower Pillow Lavas and of the Sheeted Dyke Complex are almost all metamorphosed, whereas those of the Plutonic Complex are only locally affected by ocean floor metamorphism. The lowest part of the Plutonic Complex shows a different type of metamorphism in ultrabasic rocks. The predominant rock type at this level is harzburgite, which has a foliated metamorphic texture (Moores & Vine 1971). These ultrabasic rocks are regarded as the mantle rocks underlying the ancient oceanic crust. Their metamorphism probably occurred when they were part of the Earth's mantle. Metamorphic rocks from the mantle will be discussed further in Chapter 12. They are not a product of ocean floor metamorphism, as represented in the oceanic crustal sequence in the higher part of the Troodos Complex.

The limitation of ocean floor metamorphism to the Pillow Lavas, the Sheeted Dyke Complex and the uppermost parts of the Plutonic Complex is thought to have been controlled by the depth to which sea water could percolate down through fractures in the rock at the time of metamorphism. Studies of the oxygen and hydrogen isotopes of the metamorphic rocks (Ch. 14) and of the copper deposits of the Troodos mountains both suggest that during metamorphism sea water circulated through the oceanic crust. It was heated at the deeper levels, and a complicated system of convection

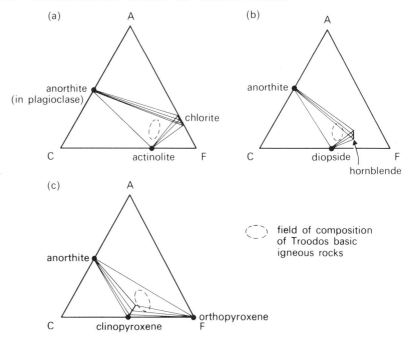

Figure 11.6 ACF triangles illustrating change of mineral assemblages with grade in basic igneous rocks of Troodos. (a) Assemblages of low-grade zone (Figure 11.3). (b) Assemblages of medium-grade zone (Figure 11.4). (c) Pyroxene-hornfels assemblage of Figure 11.5.

cells was set up. The heated H_2O in the rising parts of the convection cells would be very reactive, hydrating the surrounding basic igneous rocks. Oxygen was exchanged with the rocks and probably certain dissolved metallic ions also. The existence of this crust-wide circulation system appears to be a unique feature of metamorphism of ocean floor rocks. Although circulation systems can be shown to have formed in ground water on land also (Ch. 14), the depth of circulation of H_2O is limited to a small part of the total thickness of continental crust. This topic has only recently begun to be explored and research on these lines may fundamentally alter our view of metamorphic processes.

Since most of this chapter was written, the author has been told (J. R. Cann, personal communication) that results from the deep-sea drilling programme in the North Atlantic may shed doubt on the analogy drawn between ocean floor rocks and tectonically uplifted basic–ultrabasic complexes such as the Troodos Complex. Holes have been drilled up to 600 m into ocean floor basalts. At the time of writing, no metamorphosed basalts have been reported. The interstitial glass of the basalts is unaltered and layers of unconsolidated volcanic ash composed of basaltic glass fragments lie between basalt flows. These results obviously open up the subject for further discussion. For the moment the author supports Cann's opinion that the analogy between ocean-floor metamorphic rocks and the rocks of the Troodos Complex remains valuable for the discussion of ocean-floor metamorphism. It is possible that the lack of metamorphism in the North Atlantic basalts is due to the lack of a convective sea water circulation system at the mid-ocean ridge at the time when the ocean floor formed. More deep drilling and more heat flow measurements should help to resolve this issue.

Exercise

These analyses are of igneous rocks from the Vourinos Complex, Greece, which is an uplifted area of ocean floor rocks similar to the Troodos Complex, Cyprus. Plot the rock compositions on an ACF diagram and from this data, sketch triangles showing possible patterns of tie-lines for low-, medium- and high-grade mineral assemblages. Serpentine plots on the F corner, prehnite at A 33%, C 67%.

	(1)	(2)	(3)	(4)	(5)
SiO_2	54·7	74·7	54·7	54·56	68·4
TiO_2	0·30	0·25	0·78	0·92	0·56
Al_2O_3	16·4	12·0	14·7	16·24	13·7
Fe_2O_3	2·47	1·26	6·36	5·18	2·71
FeO	7·61	2·64	3·13	5·08	2·97
MnO	0·17	0·09	0·10	0·14	0·08
MgO	6·38	0·94	3·20	6·72	1·07
CaO	9·74	2·46	9·74	4·16	4·22
Na_2O	1·41	3·14	2·34	2·60	4·43
K_2O	0·21	0·95	0·04	0·23	0·04
P_2O_5	0·025	0·069	0·025	0·069	0·087
H_2O+	1·52	1·00	3·01	3·94	1·10
H_2O-	0·12	0·20	0·28	0·30	0·14
Total	101·05	99·70	98·41	100·09	99·51

Mineral assemblages
1. Plagioclase + hornblende + quartz + opaques.
2. Early assemblage – quartz + plagioclase + hornblende + opaques; late assemblage – quartz + epidote + albite + opaques.
3. Serpentine + prehnite + albite(?) + chlorite + opaques.
4. Prehnite + chlorite + serpentine.
5. Quartz + plagioclase + chlorite + opaques.

(From Moores (1969), with permission.)

Note Many of these rocks are oxidised. If $[Fe_2O_3] + [TiO_2]$ is greater than $[FeO]$, assume that the excess is haematite (Fe_2O_3) and therefore take F = $[MgO]$.

12

Metamorphic rocks of the upper mantle

In certain unusual geological environments rocks derived from the Earth's mantle are found at the surface. They have come to the surface in one of two ways: by tectonic movements or by volcanic eruption. Many of these rocks can be identified as metamorphic by their textures.

The density of the mantle and the rate of travel of seismic waves through it suggest that it has an ultrabasic igneous composition. This is supported by the composition of the mantle-derived rocks, which are predominantly ultrabasic. Because the metamorphism of rocks of this composition has not been discussed much in this book, ultrabasic mantle rocks will be discussed only briefly, and a more detailed description will be given of mantle-derived rocks of basic igneous composition.

The harzburgites at the base of the Troodos basic–ultrabasic complex of Cyprus have already been mentioned (Ch. 11). They are orthopyroxene peridotites with a foliation defined by preferred orientation of elongate orthopyroxene and chromian spinel crystals. Foliated dunites also occur, their foliation being due to a planar preferred orientation of olivine crystals. The foliations in these ultrabasic rocks are roughly flat-lying, parallel to the contacts with the overlying ultrabasic cumulate rocks and the ocean-floor basic igneous sequence (Fig. 11.1). It is thought that these rocks represent the solid residue of the upper mantle, left behind after partial melting. This melting gave rise to the basic magma which crystallised at a mid-ocean ridge to form the crust of the ocean floor. The foliation appears to have developed during and after the time of partial melting.

Fragments of upper mantle from beneath oceanic crust have been brought to the surface by tectonic movements (as in Cyprus) in several localities in orogenic belts, both of Alpine age and older. Mantle material from beneath both oceanic and continental crust is brought to the surface as **xenoliths** in volcanic rocks. They appear in basalts of both oceanic and continental areas, and also in the unusual volcanic rocks that are the source of diamond. These are called **kimberlites** after the famous Kimberley diamond-mining district in South Africa. The blocks in both basalts and kimberlites include rocks of basic and ultrabasic compositions. Both compositions have examples with igneous and metamorphic textures.

Kimberlites are fragmental rocks with different kinds of fragments in a fine-grained matrix. The matrix is of low-temperature minerals such as serpentine and calcite. The blocks include the country rocks seen in contact with the kimberlites and also exotic blocks which appear to have come from the deep parts of the crust and upper mantle. Kimberlites only occur in Precambrian shield areas and cratonic areas where the Precambrian shields are buried beneath Phanerozoic sediments (Fig. 2.13). They outcrop over small areas, often of irregular shape, but mining shows that the kimberlite bodies are usually vertical pipe-like structures cutting through the Precambrian rocks. Associated small dykes and sills are occasionally found. The average diameter of the pipes in the Kimberley district, which seems to be typical, is 300 m increasing gradually upwards so that the pipe has the shape of an inverted cone (Hawthorne 1975). Their structure therefore suggests that they are volcanic vents, and this supposition is supported by the discovery of erupted volcanic rocks resembling kimberlite in Tanzania and Namibia (Reid *et al.* 1975, Ferguson *et al.* 1975). The volcanic rocks form low hills which are almost certainly the eroded remains of kimberlite volcanoes.

Kimberlites are therefore considered to be volcanic rocks formed by eruptions from deep in the mantle. The material erupted at the surface was a mixture of gas and solid fragments torn from the walls of the vent. The gas probably originated in the mantle and its explosive escape upwards through mantle and crust brought rocks from very deep in the Earth to the surface. Thus the xenoliths in the kimberlite include rocks from the mantle and crust.

The mantle-derived xenoliths are of both ultrabasic and basic compositions, although, as has been stated already, the ultrabasic compositions are commoner. Both compositions may display foliation and lineation, with a considerable variety of other metamorphic textures, especially those of dynamic metamorphic rocks (Boullier & Nicholas 1975). They are much coarser-grained than crustal dynamic metamorphic rocks, with grain sizes up to ten times larger.

Figure 12.1 shows a thin section of a rock of basic igneous composition from the Roberts Victor kimberlite pipe, South Africa (Lappin & Dawson 1975, Harte & Gurney 1975). This pipe is unusual because the xenoliths in it are mainly basic in composition. In hand-specimen the two major minerals can easily be recognised as dark red garnets, up to about 10 mm across, and green pyroxene. Although very coarse-grained, the rock has an even-grained texture which is granoblastic (or granular). The microscope confirms the simple primary mineral assemblage of this rock: garnet +

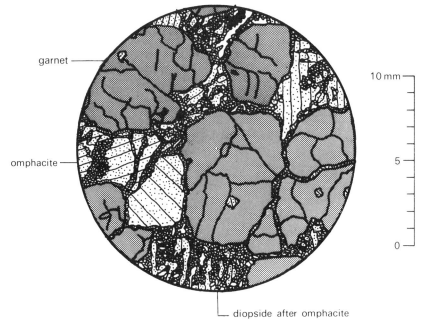

garnet

omphacite

10 mm

5

0

diopside after omphacite

Figure 12.1 Thin section of eclogite block from Roberts Victor kimberlite pipe, South Africa. Scale bar 10 mm.

clinopyroxene. Note the very low magnification of Figure 12.1. Basic igneous rocks with this mineral assemblage were mentioned in Chapter 10. They are called **eclogites**.

Eclogites have an average density of about $3 \cdot 5$ gm cm^{-3} whereas gabbros of equivalent composition have a density of about $3 \cdot 0$ gm cm^{-3}. This suggests that eclogite is a rock formed under high pressure, which caused the ions in the crystals of the minerals to adopt a closer-packed and therefore denser arrangement. This supposition has been confirmed by numerous experiments with material of basaltic composition at high pressures (e.g. Yoder & Tilley 1962). This experimental work also reveals that an eclogite assemblage may form either by solid-state transformation of basalt to eclogite, with increasing pressure at relatively low temperatures, or by crystallisation from a basaltic melt at high pressure and temperatures. The first process will give rise to a metamorphic eclogite, the second to an igneous eclogite. An eclogite with a granular texture like that in Figure 12.1 may be either a metamorphic eclogite which crystallised under stress-free conditions, or a granular igneous rock. On geochemical grounds, Harte and Gurney favour the second possibility.

At the edges of the garnet and pyroxene crystals there has been secondary alteration, which is undoubtedly metamorphic. In the rock of

Figure 12.1 the pyroxene shows more extensive retrograde metamorphism than the garnet. The large crystals of green pyroxene have been replaced along boundaries and cracks by smaller prismatic crystals. These have a similar birefringence but a lower R.I. than the large parent pyroxenes. An interesting distinction is in the optical axial angles of the two different pyroxenes. The parent pyroxenes have $2V_\gamma$ about 80°, the small secondary pyroxenes $2V_\gamma$ about 30°. This suggests that the large pyroxenes are **omphacite**, a variety of augite rich in sodium and aluminium, whereas the small recrystallised pyroxenes are more ordinary augite. Omphacite is a diagnostic mineral for eclogites and may be recognised by its green colour (although this may be too faint to be seen in thin section) and large optic axial angle. Garnet + pyroxene rocks are known in which the pyroxene is diopside or augite. These are not eclogites but garnet-bearing pyroxene gneisses (Ch. 6).

The association of eclogites with diamonds in kimberlites suggests that they formed at high pressures and therefore at considerable depths.

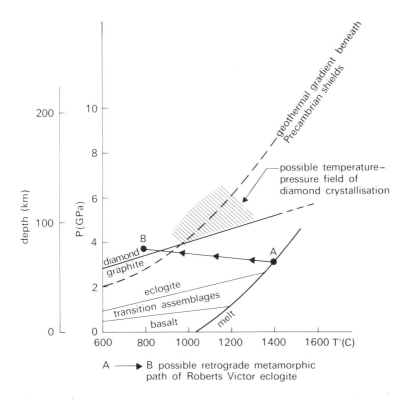

A ⟶ B possible retrograde metamorphic path of Roberts Victor eclogite

Figure 12.2   Temperature–pressure diagram showing univariant reaction curve diamond-graphite, fields of stability of basalt and eclogite and possible retrograde metamorphic history of Roberts Victor eclogites. After Lappin and Dawson (1975).

Diamonds have been found in one or two specimens of eclogite, and inclusions of eclogite and its constituent minerals have been found in diamonds, so their possible coexistence has been adequately demonstrated. Figure 12.2 shows the range of pressure and temperature over which diamond rather than graphite is the stable form of carbon. The limits of the field have been well established because of attempts to manufacture diamond industrially. The approximate depths beneath continental crust equivalent to the pressures are also given. The diagram shows that provided the assumption that diamond is a stable member of metamorphic mineral assemblages of the eclogites is correct, the metamorphism of the eclogites occurring as fragments in kimberlites must have occurred deep in the mantle.

By studying the **retrograde** changes in Roberts Victor eclogites using the electron-probe microanalyser to determine the compositions of minerals, Lappin and Dawson (1975) and Harte and Gurney (1975) have been able to trace part of the temperature and pressure history of particular xenoliths. Lappin and Dawson studied a xenolith with two different primary mineral assemblages: garnet + pyroxene and garnet + pyroxene + kyanite. Harte and Gurney studied a garnet + pyroxene xenolith with particularly well-developed exsolution lamellae of pyroxene and garnet, so that successive stages in exsolution after primary crystallisation could be studied. The metamorphic histories of these eclogites show some differences, which presumably mean that they came from somewhat different levels in the mantle when they were incorporated into the kimberlite. But there are also comparable features. During the retrograde metamorphism from the coarse-grained primary assemblages, the amount of aluminium in the pyroxene showed a steady reduction in both xenoliths. There was also in each case a remarkable change in the composition of the garnet coexisting with the pyroxene. These changes are shown on ACF triangles in Figure 12.3. The garnet field narrows down not to the almandine–pyrope composition of the garnet seen in regional metamorphic basic igneous rocks of the Barrow sequence (Fig. 8.13), but to an intermediate composition between almandine–pyrope and grossular garnets. In crustal metamorphic rocks, garnets of this composition do not occur, because of limited solid solution between garnets of the pyralspite series and garnets of the ugrandite series (Deer *et al.* 1966, p. 21). Experimental investigation of the mutual solubilities of the two garnet series indicates that garnets of these intermediate compositions are only stable in eclogites at very high pressures (an estimated $3 \cdot 0$ GPa in Lappin and Dawson's xenolith, and $3 \cdot 5$ GPa in Harte and Gurney's). Thus the retrogressive metamorphism in the eclogite xenoliths occurred by cooling at high pressures and not while the xenoliths were being erupted through the kimberlite pipe. Lappin and Dawson suggest that pressure actually increased slightly during the cooling of their xenolith.

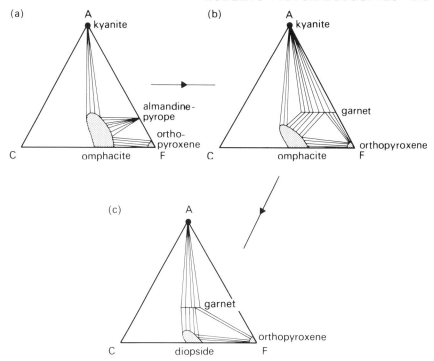

Figure 12.3 ACF diagrams showing the sequence of stable mineral assemblages in Roberts Victor eclogites. Unlike most of the sequences in this book, this is a retrograde sequence (i.e. triangle (a) represents the highest, (b) the intermediate, and (c) the lowest temperature).

Therefore it is possible to investigate metamorphic processes occurring at depths of 100–200 km in the mantle. This level is that of the low-velocity layer or **asthenosphere** in the mantle beneath continents; the exciting prospect therefore exists that in investigating metamorphic rocks from kimberlites it may be possible to determine recrystallisation processes which are important in the mechanism of continental drift, rather as Schmid's (1975) study of the Lochseiten mylonite permits discussion of the mechanism of nappe transport. Similarly study of the foliated ultrabasic rocks at the base of oceanic sequences tectonically emplaced in orogenic belts may reveal processes operating beneath oceanic lithosphere.

Finally, it should be emphasised that eclogites are not confined to mantle rocks. They also occur in rocks of the deep crust. They occur as pods, usually only a few metres across, in high-grade gneisses, especially pyroxene gneisses of the type discussed in Chapter 6. Eclogites also occur in the highest-grade parts of blueschist metamorphic rocks, again as local small blocks. These eclogites also show evidence of high pressures of metamorphism, but not as high as those of kimberlites. In the field, the density of eclogite should make it easily recognised; in thin section it should be even more distinctive.

Exercise

Analysis 1 below is the bulk composition of the eclogite block from the Roberts Victor kimberlite pipe, described by Harte and Gurney (1975). Analyses 2–6 are of garnet, 2 being from the core of the coarsest exsolution lamellae followed by a progression to 6 which is of the finest lamellae. Analyses 7–11 are of clinopyroxenes, also proceeding from the farthest removed from garnet lamellae in 7 to the clinopyroxene adjacent to the finest lamellae in analysis 11.

	(1)	(2)	(3)	(4)	(5)	(6)	(7)	(8)	(9)	(10)	(11)
SiO_2	43·50	40·70	40·84	40·21	40·82	41·28	48·33	49·04	51·43	53·82	52·20
TiO_2	0·17	0·17	0·15	0·12	0·12	0·14	0·17	0·17	0·18	0·13	0·15
Al_2O_3	15·12	22·54	22·38	22·49	22·02	21·70	13·82	11·50	7·48	5·56	7·29
Cr_2O_3	0·05	0·12	0·12	0·12	0·10	0·11	0·10	0·10	0·09	0·10	0·05
Fe_2O_3	4·68	–	–	–	–	–	–	–	–	–	–
FeO^*	4·70	10·81	9·10	9·24	9·06	8·12	4·36	4·08	3·06	3·00	2·93
MnO	0·18	0·45	0·40	0·36	0·39	0·19	0·08	0·08	0·05	0·09	0·01
MgO	12·06	11·96	11·43	9·90	9·31	7·68	10·74	12·18	13·80	13·77	13·24
CaO	17·44	14·42	15·22	17·73	18·16	21·02	22·29	23·45	22·46	22·67	21·65
Na_2O	0·64	–	–	–	–	–	1·27	1·11	1·59	1·19	1·71
K_2O	0·23	–	–	–	–	–	–	–	–	–	–
P_2O_5	0·05	–	–	–	–	–	–	–	–	–	–
H_2O+	1·22	–	–	–	–	–	–	–	–	–	–
Total	100·04	101·17	99·64	100·17	99·98	100·24	101·16	101·71	100·14	100·33	99·23

*In all analyses except 1, total iron is quoted as FeO. (Analyses by Harte and Gurney (1975) gratefully acknowledged.)

Plot all the points onto an ACF diagram. Join the corresponding garnet and clinopyroxene analyses with tie-lines. Compare the results with Figure 12.3.

Note When plotting the minerals do not subtract $[TiO_2]$ from the $[FeO]$ total, because both are of course in the garnet or pyroxene, as the case may be. For the same reason, do not subtract $[Na_2O]$ and $[K_2O]$ from the $[Al_2O_3]$ totals, in either minerals or rock.

Part III

Metamorphic rocks in the laboratory

13

The study of metamorphic reactions

In recent years, efforts to discover the conditions of formation of metamorphic rocks have concentrated upon the study of metamorphic reactions. There are two main lines of inquiry: attempts to define the chemical reactions which have occurred in rocks during metamorphism, and the study of comparable reactions in the laboratory. The preceding chapters of this book have illustrated possible chemical reactions occurring with increasing or decreasing metamorphic grade in both contact and regional metamorphism. It has been shown that chemical reactions in rocks of a particular composition are revealed by changes in the equilibrium mineral assemblage list with grade. These changes may be recorded on geological maps and cross sections as isograds. In favourable cases, the study of metamorphic reactions in the laboratory permits the temperature, pressure and chemical activity of volatile components during metamorphism to be estimated.

The laboratory study of metamorphic rocks is carried out in two ways that are often combined. Direct experiments are performed in which minerals or glasses are subjected to the temperatures and pressures under study; calculations are performed based upon the application of the laws of physical chemistry to rock materials. Both methods are usually applied to materials which are simpler chemically than metamorphic rocks; but often the laws of physical chemistry indicate that the relatively small quantities of impurities in suitable natural rocks, when compared with the materials used in experiments or calculations, will not greatly affect the numerical results. The major snag to direct experiment is that the time available for reaction in natural rocks cannot be reproduced in the laboratory. Metamorphic rocks may take millions of years to recrystallise, whilst a laboratory experiment can only run for a few months at most. Again, judicious consideration of experimental results in the light of the laws of chemistry can go some way to overcoming this problem.

The materials with simplified chemical compositions that are studied in the laboratory are referred to as **systems**, and characterised by the number and nature of their chemical components (Ch. 4). This is because experimental results are often discussed in the light of Gibbs' Phase Rule, the phases found to be in equilibrium by direct experiment or by calculation being compared with natural metamorphic mineral assemblages.

The discussion of metamorphic reactions in this chapter is selective, concentrating upon reactions which have been encountered in the equally selective petrographic and field descriptions of metamorphic rocks in the earlier chapters. For fuller surveys of the possible reactions in metamorphic rocks of several different compositions the reader is referred to Miyashiro (1973) and Winkler (1976). The application of the laws of physical chemistry to metamorphic petrogenesis is best described in Turner (1968, Chs 3 and 4).

An important distinction needs to be drawn between two categories of metamorphic reaction. There are differences in experimental technique and in the type of calculation employed between solid–solid reactions and reactions in which volatile components, such as H_2O or CO_2, participate. An example of the first type of reaction is:

$$NaAlSi_3O_8 \rightleftharpoons NaAlSi_2O_6 + SiO_2$$

albite $\rightleftharpoons$ jadeite $+$ quartz

In such solid–solid reactions, the position of the univariant curve on a temperature–pressure diagram is independent of the presence or absence of a volatile phase during metamorphism. An example of the second type of reaction is:

$$KAl_3Si_3O_{10}(OH)_2 + SiO_2 \rightleftharpoons KAlSi_3O_8 + Al_2SiO_5 + H_2O$$

muscovite $+$ quartz $\rightleftharpoons$ K-feldspar $+$ andalusite $+$ water vapour

In reactions like this in which a volatile phase participates the chemical activity of the component H_2O in the volatile phase plays a large part in controlling the shape of the univariant reaction curve on a temperature–pressure diagram. Because of the technical problems of supplying a volatile phase at the temperatures and pressures of interest in metamorphic petrology, there may also be considerable differences in experimental apparatus and technique for the two types of reaction.

DIRECT EXPERIMENT

It is now possible to achieve temperatures and pressures in laboratory apparatus equivalent to the entire range of conditions which are likely to be present in the Earth's crust and in a considerable part of the mantle also.

The apparatus used is elaborate and expensive and requires skilled and experienced people to operate it and to interpret the experimental results. As a result work of this kind is confined to a few laboratories, of which the Geophysical Laboratory of the Carnegie Institute, Washington, USA, is the oldest and most famous.

The system Al₂SiO₅

Because three polymorphs of Al_2SiO_5 occur widely in rocks of pelitic composition and inversion of one polymorph to another has apparently occurred in many cases (Chs 8, 9), this system has been widely studied experimentally. Three solid–solid reactions are involved, all in phases with the same composition, Al_2SiO_5. This is therefore a one-component system, which should make it especially easy to investigate and to apply the results. The three solid–solid reactions are:

$$\text{andalusite} \rightleftharpoons \text{kyanite}$$
$$\text{andalusite} \rightleftharpoons \text{sillimanite}$$
$$\text{sillimanite} \rightleftharpoons \text{kyanite}$$

The temperature–pressure diagram showing the three univariant curves for

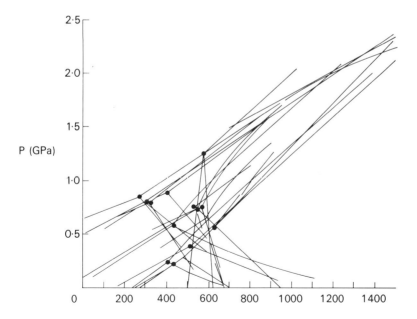

Figure 13.1 The Al_2SiO_5 reaction curves in positions inferred by various experimenters. From Vernon (1976).

these reactions should also contain a triple point, marking a unique temperature and pressure at which andalusite, kyanite and sillimanite are stable together.

The results of a large number of experiments by different laboratories are disappointingly varied (Fig. 13.1). Some of the variation can be explained

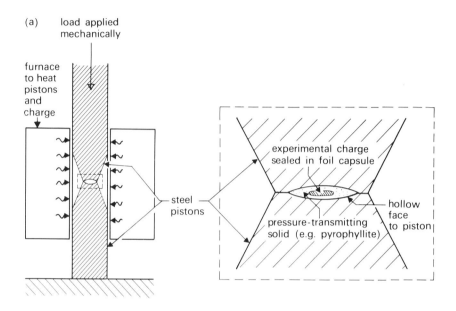

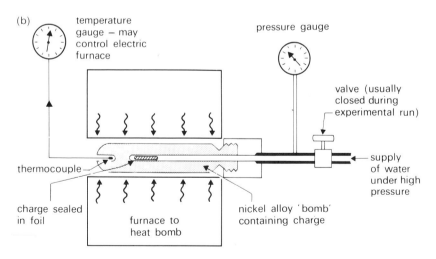

Figure 13.2 (a) Diagram of a 'simple squeezer' apparatus. (b) Diagram of hydrothermal apparatus.

by experimental technique. Early attempts to determine the positions of the univariant curves used a simple squeezer apparatus or similar instrument (Fig. 13.2a). In such an apparatus, the pressure in the experimental capsule was not measured directly but calculated from the load on the piston. It has since been shown that these calculations were wrong because the pressure within the capsule varied from place to place and was on average lower than the calculations had indicated. Because of the bad results from this type of apparatus, W. S. Fyfe has called them 'abominable squeezers'.

More recently, experiments have been performed using hydrothermal apparatus (Fig. 13.2b). The experiments are usually performed by trying to grow crystals of one polymorph of Al_2SiO_5 from another. The mineral is ground to a fine powder to increase its surface area as much as possible, and is in contact with supercritical H_2O vapour at the appropriate temperature and pressure. Crystals of andalusite, kyanite and sillimanite will not grow spontaneously in these circumstances. The charge must be seeded with natural crystals of one or more polymorphs. It is then usually possible to show that some of the small seeds have grown during the experimental run, while others have dissolved slightly. Even following this procedure, runs under temperatures and pressures close to a univariant line or the triple point may give ambiguous results. In spite of all these difficulties, results have now been obtained which are consistent with one another and with results obtained by thermodynamic calculation. The curves generally regarded as most acceptable are shown in Figure 13.3, after Richardson *et al.* (1969). For reasons to be explained in the next section, reactions from one polymorph to another are unusually slow.

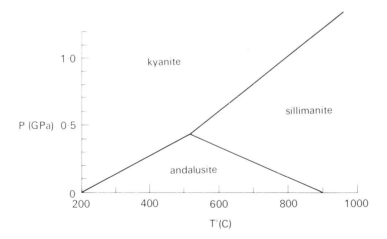

Figure 13.3 Stability relationships of andalusite, kyanite and sillimanite in the Al_2SiO_5 system. From Vernon (1976) after Richardson *et al.* (1969).

Because of the great difficulty in the experiments and the variety of results some workers (e.g. Winkler 1976, pp. 91–94) have expressed reservations about the applicability of experiments in the Al_2SiO_5 system to the estimation of temperatures and pressures in metamorphosed pelitic rocks. The possibility that the curves may be affected appreciably by the presence of other ions during metamorphism, especially Fe^{+3}, supports this view. The circumstances of growth of the polymorphs in experiments, from one powdered polymorph onto seeds of another, are not very like those under which the polymorphs grew in nature. However, to reject the curves altogether is certainly too pessimistic, because they fit in quite well with other less uncertain experimental results on mineral stability. The temperature obtained for the univariant point also fits well with the results of oxygen isotope geothermometry (Ch. 14).

The breakdown of muscovite

It was shown in Chapter 5 that in the inner part of the Comrie Aureole the muscovite of the country rocks has broken down, giving mineral assemblages containing potash feldspar. The reactions involved are dehydration reactions, and similar reactions have been studied in the system $KAlSi_3O_8 - SiO_2 - Al_2O_3 - H_2O$ (a part of the system $K_2O - SiO_2 - Al_2O_3 - H_2O$).

Two reactions occur in this simple system which are similar to the reactions that occur in pelitic rocks. One is the breakdown of muscovite when the system is over-saturated with SiO_2. This reaction is expressed as the equation:

$$KAl_2(AlSi_3O_{10})(OH)_2 + SiO_2 \rightleftharpoons KAlSi_3O_8 + Al_2SiO_5 + H_2O \qquad (1)$$

$$\text{muscovite} + \text{quartz} \rightleftharpoons \text{K-feldspar} + \text{andalusite} + \text{water vapour}$$

When quartz is not present in excess in the experimental system, muscovite breaks down as follows:

$$KAl_2(AlSi_3O_{10})(OH)_2 \rightleftharpoons KAlSi_3O_8 + Al_2O_3 + H_2O \qquad (2)$$

$$\text{muscovite} \rightleftharpoons \text{K-feldspar} + \text{corundum} + \text{water vapour}$$

The experimental curves for these reactions are shown in Figure 13.4.

Both reaction (1) and reaction (2) are dehydration reactions. They are very sensitive to the activity of H_2O in the vapour phase, and the curves in Figure 13.4 were determined by hydrothermal experiment, where the activity of H_2O would be very high. It is not certain whether in nature, for example in the Comrie Aureole, the condition during the breakdown of muscovite was the same. If the rocks of the contact aureole always had a small excess in their water content over the H_2O contained in the crystal

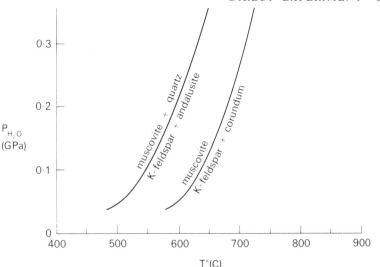

Figure 13.4 Univariant curves for the breakdown of muscovite. From Turner (1968).

structures of hydrous minerals such as muscovite, and progressive metamorphism involved a continuous driving out of H_2O from the rocks, the activity would be high and the curves would represent the temperatures and pressures of metamorphism quite accurately. This is how petrologists imagine that the progressive metamorphism of initially water-rich sediments like pelitic rocks occurred. But for rocks which were initially dry, such as basic igneous rocks, the activity of H_2O could have been small during much of the metamorphism. For carbonates, from which CO_2 as well as H_2O was driven out during metamorphism, the ratio of these components in the volatile phase driven out would affect the metamorphic sequence, as described in Chapter 5.

The shape of the dehydration breakdown curves of muscovite in Figure 13.4 shows a positive slope, increasing with increasing pressure. This is the characteristic shape of dehydration curves and is due to the fact that the volume of the products on the right-hand side of equations (1) and (2) is much larger than the volume of those on the left, especially at low pressures. The reason for the increase in volume from left to right is, of course, the presence of the water vapour on the right. Temperature–P_{CO_2} curves for decarbonation reactions in systems with CO_2 as a reactant, but without H_2O, are the same shape. These reactions do not present quite such acute problems of sluggish reaction as those of the Al_2SiO_5 system.

Provided that the temperatures and pressures of the invariant curves for the Al_2SiO_5 system and the $KAlSi_3O_8$–Al_2O_3–SiO_2–H_2O system are not

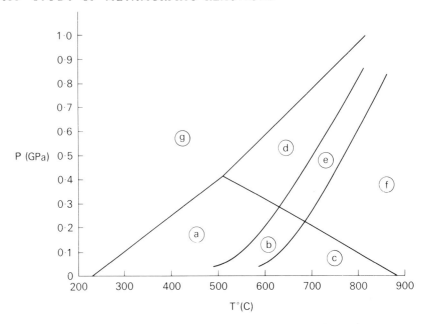

Figure 13.5 Superimposition of Figures 13.3 and 13.4 to make a simple petrogenetic grid. Circled letters refer to triangles of Figure 13.6.

affected by the presence of small amounts of Na_2O, FeO, MgO, etc., it is possible to estimate fields of temperature and pressure over which particular assemblages will be stable, for pelitic rocks which are only rich in K_2O, Al_2O_3, and SiO_2. The fields of stability are shown on Figure 13.5 and small triangular diagrams representing the stable mineral assemblages are shown in Figure 13.6.

If only rocks with compositions close to this simplified four-component system are considered, it could be said that the Comrie Aureole shows the sequence:

$$(a) \rightarrow (b) \rightarrow (c)$$

The contact aureole of the Sulitjelma Gabbro, Norway (Fig. 8.1), on the other hand, shows a different sequence. The country rock surrounding the gabbro is high-grade schist containing kyanite. In the aureole this breaks down to sillimanite. Muscovite is present in the country schists, but the innermost aureole contains potash feldspar. The sequence in this contact aureole is therefore:

$$(g) \rightarrow (d) \rightarrow (e) \rightarrow (f)$$

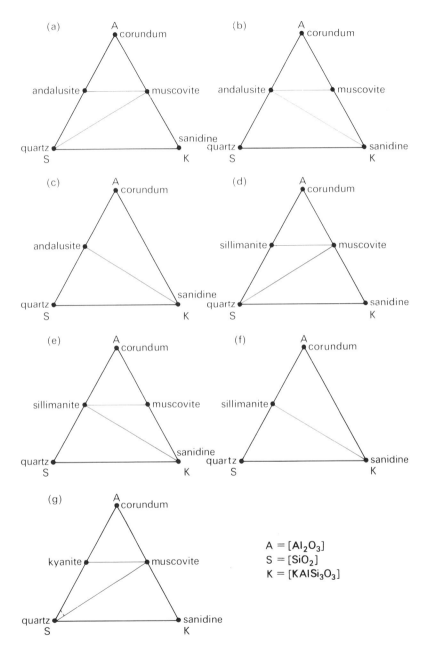

Figure 13.6 Triangular diagrams showing mineral assemblages stable within the numbered temperature–pressure fields of Figure 13.5.

Under-saturated assemblages containing potash feldspar occur, justifying the inclusion of field (6), although it is likely that in the inner aureole at Sulitjelma, the activity of H_2O might have been low and the muscovite breakdown curves might not be reliable.

The simple system therefore illustrates clearly that the progressive contact metamorphism at Sulitjelma occurred at higher pressures than that at Comrie or Skiddaw.

Figure 13.5, combining the results of experiments in two different systems and using these to discuss the mineral assemblages of actual rocks, is a very simple version of a **petrogenetic grid**. The mineral assemblages shown on the small triangles of Figure 13.6 are each in equilibrium over a restricted range of temperature and pressure, shown on Figure 13.5. Thus mineral assemblages may be used to determine temperatures and pressures of metamorphism. It is too simple for serious petrogenetic discussion because it ignores too many of the chemical components (e.g. MgO and FeO) in the pelitic rocks being studied. Experimental work has been done on more complex systems, however, and has been used as the basis of more elaborate petrogenetic grids that are applicable to a wider range of rock compositions (e.g. Naggar & Atherton 1970, Harte 1975). As more reactions are considered, the number of fields increases and, theoretically, the field of temperature and pressure for each combination of mineral assemblages can be narrowed down greatly. But when substitution of one ion for another in the same mineral species (e.g. $Mg^{+2} \rightleftharpoons Fe^{+2}$ in biotite) is possible, a different type of reaction becomes possible and further complications arise.

Sliding reactions

In both the simple experimental systems discussed so far, the minerals have fixed chemical compositions and the metamorphic reactions considered result in abrupt changes in the mineral assemblages. Reactions of this type are called **discontinuous reactions**. In many minerals, including such common ones in metamorphic rocks as garnet, hornblende and biotite, metallic ions of similar size may substitute for one another. Probably the substitution most studied in metamorphic rocks is that between Mg^{+2} and Fe^{+2} mentioned above. There are others which are also important. Pairs of ions may substitute for other pairs (e.g. $Na^{+1}Al^{+3}$ for $Ca^{+2}Mg^{+2}$).

Because of the complexity of sliding reactions, few have been exhaustively investigated experimentally. One set of reactions which have been studied in some detail are those in the system $MgO-FeO-Al_2O_3-SiO_2-H_2O$ which define the stability of cordierite and almandine–pyrope garnet. One reaction from quite a number of possibilities will be considered here: a change with increasing pressure from the assemblage cordierite + hypersthene + quartz to the assemblage garnet + sillimanite + quartz. The results

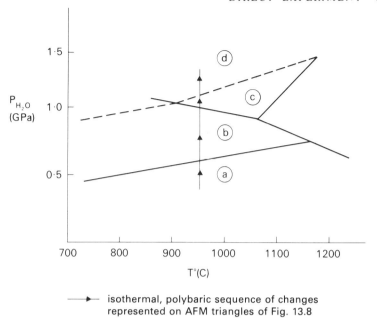

⟶ isothermal, polybaric sequence of changes represented on AFM triangles of Fig. 13.8

Figure 13.7 Temperature–pressure ranges of stability of certain mineral assemblages for a system whose composition is represented on the AFM triangles of Figure 13.8. From Hensen and Green (1970).

discussed here are from experimental work by Hensen and Green (1970) using one particular composition within the four-component system $MgO–FeO–Al_2O_3–SiO_2$. Because H_2O does not appear as a reactant in these experiments, the reactions are solid–solid reactions. In this account, therefore, the role of H_2O will not be considered although it is not strictly correct to ignore its activity (Wood 1973). The errors introduced by making this assumption may affect precise estimates of temperature and pressure based upon compositions of coexisting cordierite and garnet in natural rocks, but they do not alter the principles expounded here. Quartz is assumed to be present in all the assemblages under consideration and there is no K_2O in the system. The minerals may therefore be represented directly on AFM triangles, without the projection from muscovite composition used in the Thompson AFM diagram.

The reaction by which garnet is formed at the expense of cordierite and hypersthene in field b of Figure 13.7 may be written as follows:

$$Al_3(Mg,Fe^{+2})_2(Si_5AlO_{18}) \quad + \quad (Mg,Fe^{+2})SiO_3$$

cordierite hypersthene

(3)

$$\rightleftharpoons \quad 2(Fe^{+2},Mg)_3Al_2Si_3O_{12} \quad + \quad SiO_2$$

garnet quartz

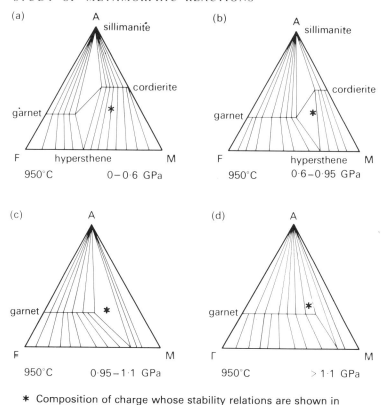

(a) A sillimanite, cordierite, garnet, hypersthene M, F, 950°C, 0–0·6 GPa

(b) A sillimanite, cordierite, garnet, hypersthene M, F, 950°C, 0·6–0·95 GPa

(c) A garnet M, F, 950°C, 0·95–1·1 GPa

(d) A garnet M, Γ, 950°C, > 1·1 GPa

∗ Composition of charge whose stability relations are shown in Fig. 13.7

Figure 13.8 AFM triangular diagram showing the composition of the experimental charges from which Figure 13.7 is derived, and the stable mineral assemblages (+ quartz in all cases) in fields a, b, c and d of Figure 13.7.

The reaction starts to occur at 950 °C and 0·6 GPa in the particular composition within the four-component system which is being considered (Fig. 13.8). At 950 °C, between 0·6 and 0·95 GPa, cordierite, hypersthene, garnet and quartz coexist (Fig. 13.8b). At first the garnet is relatively rich in Fe^{+2}, as the formulae written in equation 3 indicate. As the pressure increases the proportion of garnet in the mineral assemblage increases and the proportions of cordierite and hypersthene fall. The compositions of all three minerals change. As the garnet becomes richer in magnesium, its proportion increases. As the cordierite and hypersthene also become more magnesium-rich, their proportion decreases. These changes are illustrated in the triangles of Figure 13.8a to b. For the composition in the system being discussed, the reaction does not continue until all the cordierite is used up.

At 0·95 GPa another reaction occurs, which eliminates the remaining cordierite:

$$Al_3(Mg,Fe^{+2})_2(Si_5AlO_{18}) + (Fe^{+2},Mg)_3Al_2Si_3O_{12} + (Mg,Fe^{+2})SiO_3$$

cordierite   garnet 1   hypersthene 1

(4)

$$\rightleftharpoons (Mg,Fe^{+2})_3Al_2Si_3O_{12} + 3(Mg,Fe^{+2})SiO_3 + 2Al_2SiO_5 + SiO_2$$

garnet 2   hypersthene 2   sillimanite   quartz

This is an even more complex reaction, sliding with respect to some minerals and chemical components and not to others. It will not be discussed further here, except that it is illustrated cutting off the pressure range of reaction (3) in Figure 13.7, and is shown by the change from triangle b to c in Figure 13.8.

Sliding reaction (3) is comparatively simple. The reaction may be calibrated by determination of the compositions of the coexisting cordierite, hypersthene and garnet formed in the experiments. It would then be possible to determine the temperature and pressure of metamorphism in rocks with a similar simple composition, subject to the qualification about making corrections for the activity of H_2O mentioned earlier. The temperature and pressure could be uniquely determined by this method, not merely fixed within a range of pressure and temperature. The electron probe micro-analyser (Ch. 15) is capable of performing the necessary mineral analyses both of the experimental materials and of corresponding natural minerals.

This type of study has a considerable potential for determination of temperatures and pressures of metamorphism. The problems have been briefly touched upon. The petrographic problem is the establishment of a **realistic reaction** (Vernon 1976, Ch. 4) by study of changing mineral assemblages and of metamorphic textures. This is not easy; many reactions have been discussed which subsequent careful study of the rocks has shown to be simplistic. The experimental problem is to control all likely variable factors. The importance of the chemical activity of H_2O has been mentioned. In the reaction described above, Fe^{+2} is an important participant. This means that the oxidation state of the rocks has to be considered. The reaction would yield different ratios of Mg to Fe in the minerals if Fe^{+2} were undergoing oxidation to Fe^{+3} while it was going on. Hensen and Green consider that in their experimental apparatus, which was made of steel, it is unlikely that oxidation could occur. It is possible to control accurately the amount of oxidation in an experimental system, but this introduces another factor to be controlled in what are already lengthy complicated experiments.

Many of the chemical reactions which occur at isograds, such as the garnet-forming reaction at the isograd in Sulitjelma (Ch. 8) are sliding reactions. Often, early guesses at the reaction occurring at an isograd are that it is a **discontinuous** reaction, but subsequent more careful studies show that it is a sliding reaction (Atherton 1968). Potentially, an accurate knowledge of rock and mineral compositions on either side of the isograd, combined with experimental study of the realistic reaction could yield accurate values of the temperature, pressure, activity of H_2O, degree of oxidation and so forth which obtained during metamorphism. The labour involved in such a study would be very great.

The definition of the reactions occurring at isograds is often the aim in modern studies of regional metamorphic rocks. In classical regional metamorphic sequences, such as the Barrow zones of the Grampian Highlands of Scotland, considerable progress has been made with this task. There is a good deal of difference of opinion between different researchers. These efforts may yield a consistent pattern of particular reactions which can be applied generally to rocks of a particular composition (e.g. pelitic rocks and basic igneous rocks), but in the author's opinion it is too early to do this at present. For this reason, in this book, a descriptive definition of isograds has been preferred to a definition framed in terms of metamorphic reactions as used by Winkler (1976) for example.

CALCULATIONS OF MINERAL STABILITY FIELDS

An alternative to direct experiment is the calculation of mineral stability ranges of temperature and pressure. Usually, the two techniques are used together, experiment providing numerical information on which calculations may be based.

Two kinds of calculation are of particular interest for the study of metamorphic reactions. The first is the determination of equilibrium relationships in simplified systems – the theoretical equivalent of the experiments described in the preceding section. The second is the study of the rate of metamorphic reactions, a subject known as reaction kinetics. Calculation has an advantage over experiment in this field (although some direct experiments have been performed) because of the short time available for experiments compared with geological processes. Experiment is still needed to supplement calculation in this field because the steps of a reaction can only be determined by experiment. For studies of reaction kinetics it is necessary to know the equilibrium state towards which reaction is proceeding. Therefore the calculation of equilibrium relationships will be discussed first.

The simple sedimentary to metamorphic rock transition discussed at the beginning of Chapter 4 will be used again as an example. The reader will remember that two minerals A and B present together in equilibrium in the sedimentary rock, became unstable together under metamorphic conditions. In other words, the metamorphic reaction

$$A + B \rightleftharpoons C$$

proceeded from left to right. The equilibrium assemblage under the conditions of metamorphism was $B + C$ because there was an excess of B in the starting material. From the discussion in the earlier part of this chapter it will be clear that this is a discontinuous solid–solid reaction.

The reaction goes from left to right because under the temperature and pressure conditions of metamorphism the chemical potential energy of $A + B$ is greater than that of C. The recrystallising rock will tend to form the lowest energy mineral assemblage, that is the equilibrium mineral assemblage. If $A + B$ is stable in the sedimentary rock, then under those conditions the energy of those two minerals is lower:

Low T and P
energy of A + energy of B < energy of C
High T and P
energy of A + energy of B > energy of C

The chemical potential energies must be determined for the proportions of A and B which actually participate in the reaction. For example, consider the solid–solid reaction:

$NaAlSi_2O_6$	$+ SiO_2$	$\rightleftharpoons NaAlSi_3O_8$
jadeite	quartz	albite
$202 \cdot 04$ g	$60 \cdot 07$ g	$262 \cdot 11$ g

Possible weights for the quantities of the reactants are given. It is usual to calculate energies for the weights given above, which are the chemical molecular weights in grams, based upon the chemical formulae of the minerals used in the equations. These weights are called **gram molecular weights** usually abbreviated to **moles**. The reaction could be written:

1 mole jadeite + 1 mole quartz $\rightleftharpoons$ 1 mole albite

Returning to the hypothetical example, therefore, under the high temperature and pressure conditions, if the reaction is to proceed from left to right:

energy of 1 mole A + energy of 1 mole B > energy of 1 mole C   (5)

The chemical energy of a mineral at a particular temperature and pressure may be calculated from knowledge of such properties as its energy of formation from its constituent elements and the variation of its specific heat with temperature. These properties are not easy to measure; but in some cases satisfactory agreement with direct experiment and petrographic experience has been obtained. The method is much more successful in extending limited parts of a univariant curve for a discontinuous reaction obtained by experiment or, for extending over a wider field, values for the compositions of coexisting minerals in sliding reactions. The combination of experiment and calculation is more effective for providing a basis for the determination of the conditions of metamorphism than direct experiment alone. Calculation can check the experimental results for consistency and recognise the effects of unexpected reactions as well as extrapolating results. It is likely that this approach will be more extensively employed in future. Wood and Fraser (1977) give a full and rigorous treatment of the subject.

In the calculation of reaction rates, what is important is the *change* of energy as the reaction proceeds in one direction or the other. If the change is large, the reaction will tend to be quick; if the change is small it will tend to be slow. The chemical potential energy of minerals is expressed as a thermodynamic parameter known as **Gibbs' free energy** and given the symbol G. Thus the inequality expression (5) given above could be written simply as

$$G_A + G_B > G_C \tag{6}$$

The change of energy during this reaction is given the symbol ΔG. For (6) above, it follows that

$$\Delta G = (G_A + G_B) - G_C \tag{7}$$

ΔG may be discovered for some reactions by making calorimetric measurements during experimental runs; or it may be calculated. Fyfe *et al.* (1958) long ago pointed out that for many reactions of interest in metamorphic petrology, ΔG is small compared with the total energy of the participating minerals. They called this the 'plague of the small ΔGs' (Fyfe *et al.* 1958, pp. 22–25). This is true, for example, of the polymorphic transitions between andalusite, kyanite and sillimanite, which explains why they are so sluggish, particularly near the univariant curves and the invariant triple point.

The detailed study of reaction rates is a complex topic, because it includes discussion of the mechanisms of mineral nucleation and growth in the solid state, and also a considerable discussion of diffusion of metal ions through crystal structures and along grain boundaries; it lies beyond the scope of this book. It is treated in Vernon (1976).

Topological analysis of petrogenetic grids

A method which has recently been applied successfully in the preparation of petrogenetic grids uses the geometrical constraints which are a consequence of the Phase Rule (Ch. 4). It works best where a number of discontinuous reactions occur in systems with several chemical components. The method was invented by Schreinemakers in the 1920s, but its application to metamorphic petrology is new. It provides a check upon the consistency of the results of direct experiments and permits an extrapolation of temperature–pressure fields of stability of particular assemblages to parts of petrogenetic grids where direct experimental results are not available. The method has been applied, for example, by Harte (1975) to construct a petrogenetic grid for pelitic rocks in the Barrow metamorphic sequence of the Grampian Highlands. He assumed that in this classic area the activity of H_2O was high during metamorphism and that oxidation did not occur during the metamorphism of silicate minerals containing Fe^{+2}. He found that his model predicted well the mineral assemblages actually found. This is encouraging for the applicability of experimental results to regional metamorphic rocks generally. He therefore concluded that the petrogenetic grid he computed represents reliably the maximum temperatures and pressures achieved during metamorphism, which varied in different parts of the area. It is often assumed that temperature–pressure grids based upon experimental studies do this, but the underlying assumptions are seldom so rigorously spelt out and checked.

The details of Schreinemakers' technique for constructing petrogenetic grids are not given here, but may be found in Miyashiro (1973, pp. 133–135) or Vernon (1976, pp. 44–45).

CONCLUSIONS

The estimation of temperatures and pressures achieved during metamorphism depends upon the identification of mineral assemblages and the determination of metamorphic reactions. The precision of the estimates will depend upon the reaction concerned and upon the amount and quality of the experimental data about it. It is particularly hard for geologists without experience of experimental work themselves (like the author) to form judgements on the quality of experimental results. The achievements of experiment, for example in demonstrating the high pressures of formation of blueschists and eclogites, have enabled the results of metamorphic petrology to be incorporated into the wider understanding of the evolution of the Earth. This is an immense advance over the state of affairs when Alfred Harker wrote his classic textbook on metamorphism in 1932. But conflicting and inaccurate experimental results, often presented with no hint of uncertainty, must produce a certain scepticism in the non-specialist. The following paragraph tries to adopt a balanced view, within the considerable limitations of the author's knowledge.

In suitable cases, it appears that temperatures can be estimated remarkably accurately ($< \pm 5°C$). More often they can be fixed within limits of 20–50°C. Pressure estimates are less reliable, with errors seldom less than 0·1 GPa, although relative pressures can often be more reliably determined than absolute values. Pressure estimates also depend upon assumptions about the activity of H_2O and other volatile components. The reader will appreciate that temperature and pressure estimates for the metamorphism of natural rocks rest on a large number of assumptions, e.g. that the minerals studied were in equilibrium during metamorphism, or that the experimental results used are reliable and about the activities of volatile components. In accounts of particular metamorphic rocks, research workers are rarely able to provide rigorous checks on all of them, experimentalists taking the interpretation of field relationships and textures on trust, field petrologists taking experimental results on trust. As a field petrologist, the author would particularly welcome a comprehensive and critical review of the 'state of the art' in experimental petrology as applied to metamorphic rocks.

METAMORPHIC FACIES

The idea of metamorphic facies was proposed by the great Finnish petrologist, P. Eskola, in 1915 before there were any significant experimental or thermodynamic data on the stability of metamorphic minerals. In its recent development the idea has become inseparable from research on mineral stability. In this form its fullest and most rigorous exposition is to be found in Sobolev (1972). The facies approach to the study of metamorphic rocks has not been followed in this book, for reasons to be explained later. But the student will encounter frequent references to metamorphic facies in the literature of geology, including many of the works cited in this book, and therefore an outline is presented in this chapter.

Metamorphic facies takes the recognition of mineral assemblages a stage further. A mineral assemblage in a rock of a particular composition is only stable over a finite range of metamorphic conditions. Eskola's scheme of 1920 listed a number of diagnostic mineral assemblages in metamorphosed basic igneous rocks, which define broad fields of temperature and pressure, in the same way that the simplified assemblages shown in Figure 13.6 define fields on the petrogenetic grid in Figure 13.5. The only factors which this early scheme considered to be varying during metamorphism were temperature and pressure; five different ranges were distinguished and given the name **metamorphic facies**. Later authors have considered wider ranges of rock composition and more variables (such as the activities of H_2O and CO_2) and have therefore generally proposed a greater number of metamorphic facies.

It can be seen that metamorphic facies is a quasi-genetic classification of the conditions of metamorphism of rocks. It is *quasi*-genetic, not strictly genetic, because it is not necessary to know the numerical values of temperature, pressure and activities of volatile components in order to define a particular metamorphic facies. Individual facies are given names after common rock types stable under the appropriate conditions, and in schemes with large numbers of subdivisions may be divided into sub-facies, which are usually called after mineral assemblages.

An individual metamorphic facies may be defined in one of two ways. Critical mineral assemblages may be defined for particular rock compositions (e.g. assemblages with glaucophane + lawsonite in basic igneous rocks of the blueschist facies). This was the 'classical' method of definition of Eskola and has often been followed since. Alternatively, critical metamorphic reactions may be taken to define the boundaries of metamorphic facies (e.g. muscovite + quartz $\rightleftharpoons$ potash feldspar + H_2O for the amphibolite to pyroxene hornfels facies boundary). This approach is popular today because metamorphic reactions are being studied.

As in all classification schemes there is a dispute between those who propose more and more subdivisions as more is discovered about metamorphic rocks ('splitters') and those who prefer to retain broad divisions, but have to modify the definition of facies as they go along ('lumpers'). Modern studies of metamorphic rocks have been guided by the very clear exposition of metamorphic facies and its foundations in experimental petrology given by Fyfe, Turner and Verhoogan (1958). This exposition was enlarged by Turner (1968). Also very influential, expounding a slightly different facies scheme, was Winkler (1965). Because of the influence of these books on teaching and research, the descriptions of metamorphic rocks in many scientific papers published in the decade of geological discovery from 1965 to 1975 are couched in the language of metamorphic facies.

The approach to the description of metamorphic rocks in terms of metamorphic facies has some important advantages. It encourages the recognition of *all* the minerals in metamorphic rocks, in order to obtain assemblage lists. It also encourages the study of rocks of several different compositions, if possible, before pronouncements are made about the conditions of metamorphism. It is of wide applicability and particularly useful for the description of metamorphic rocks, such as blueschists, formed under extreme metamorphic conditions.

The definition of metamorphic facies, both individually and collectively, has been a problem ever since Eskola first introduced the idea. There are uncertainties in recognising the reactions limiting metamorphic facies fields, and in fixing the temperature–pressure curves for these reactions. These uncertainties of determination of temperature and pressure of facies boundaries may be comparable to the temperature and pressures ranges of the facies themselves, even the broad divisions of the 'lumpers'. It is this

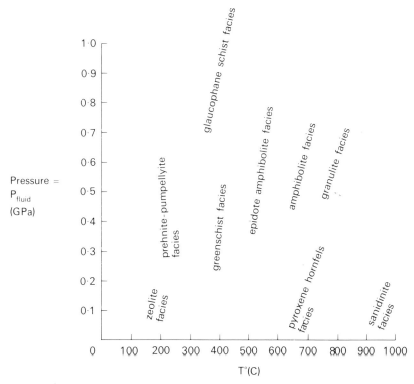

Figure 13.9 Approximate positions of metamorphic facies on a temperature–pressure diagram. After Miyashiro (1973).

problem which has led Winkler (1976) and others to abandon the facies approach. Miyashiro (1973) avoids the difficulty by adopting broad temperature–pressure fields for individual metamorphic facies and concentrating upon description of progressive metamorphic sequences in metamorphic rocks, rather than becoming involved in detailed discussion of facies boundaries or diagnostic mineral assemblages. Thus his scheme has practical value for general descriptions of the whole range of conditions of metamorphism and metamorphic rocks, but it could be argued that he begs the theoretical question of definition.

Miyashiro's scheme will be briefly expounded here, to help the student to understand the geological literature. It will be appreciated that Miyashiro is a 'lumper'. His broad scheme means that the only factors considered are temperature and pressure, as in Eskola's original scheme. This does not mean that other factors are not considered in Miyashiro's book, merely that his facies scheme disregards them. The assumption is that the activity of H_2O is high during metamorphism, as in hydrothermal experiments. Figure 13.9 shows the fields of the ten metamorphic facies which

Table 13.1 Names of metamorphic facies and typical mineral assemblages of basic igneous rocks and pelitic rocks in each.

Facies name	Typical assemblage in basic igneous rocks	Typical assemblage in pelitic rocks
greenschist	chlorite + actinolite + albite + epidote + quartz	chlorite + muscovite + chloritoid + quartz
epidote- amphibolite	hornblende + epidote + albite + almandine garnet + quartz	almandine garnet + chlorite + muscovite + biotite + quartz
amphibolite	hornblende + andesine + garnet + quartz	garnet + biotite + muscovite + sillimanite + quartz
pyroxene hornfels	clinopyroxene + labradorite + quartz	cordierite + andalusite + biotite + quartz
granulite	clinopyroxene + labradorite + orthopyroxene + quartz	garnet + cordierite + biotite + sillimanite + quartz
sanidinite	clinopyroxene + labradorite + quartz	sanidine + sillimanite + hypersthene + cordierite + quartz
glaucophane schist	glaucophane + lawsonite + quartz	muscovite + chlorite + spessartine garnet + quartz
eclogite	pyrope-garnet + omphacite	not known
zeolite	smectite + zeolites (1) (with relict igneous plagioclase and clinopyroxene)	illite + chlorite (2) + quartz
prehnite- pumpellyite	prehnite + (1) pumpellyite (with relict igneous plagioclase and clinopyroxene)	illite + stilpnomelane (2) + chlorite + quartz

(1) Basic igneous rocks of the zeolite and prehnite-pumpellyite facies seldom have an equilibrium mineral assemblage. Common relict minerals of the igneous rocks are therefore given in brackets.

(2) The zeolite and prehnite-pumpellyite facies have been defined in metamorphosed grey-wackes and basic igneous rocks, but not in pelitic rocks. The pelitic assemblages given here are guesses by the author.

Table based upon Miyashiro (1973) especially Chapter 11.

Miyashiro distinguishes; the metamorphic assemblages for each facies are tabulated in Table 13.1 for basic igneous rocks and for pelitic sediments. Since the definition of diagnostic mineral assemblages is such a problem, it should be emphasised that the assemblages given are not necessarily diagnostic, though the author hopes they are typical. They may refer to slightly different basic igneous and pelitic sedimentary rock compositions in each facies.

The reader may wonder why a metamorphic facies scheme has not been adopted right through this book, particularly as such a strong emphasis has been placed upon the recognition of equilibrium mineral assemblages. The reason is that the book is intended for students at a stage when they should be learning to recognise minerals and describe metamorphic rocks accurately for themselves. Assignment of a suite of metamorphic rocks to a metamorphic facies should come at the end of an extensive programme of field and microscopic study of metamorphic rocks, not at the beginning. The author has been disturbed by an increasing tendency among students to arrive prematurely at a metamorphic facies name for a suite of rocks. This can lead to serious mistakes, such as failure to recognise minerals not associated with the facies or, worse, incorrect identification of minerals which 'ought' to be present. Examples of the last error in the author's direct experience are clinozoisite described as andalusite in a contact aureole, and epidote described as lawsonite in a glaucophane-bearing schist. This book has therefore deliberately not introduced a metamorphic facies scheme and fitted the rocks into the appropriate pigeon-holes in it. Instead, field relations and a loose definition of metamorphic grade have been used to provide a framework for discussion. It is hoped that the loss in clarity of exposition is matched by a gain in appreciation of the uncertainty of conclusions about metamorphic petrogenesis and the need for them to be soundly based in accurate observation.

However, it would not be fair for the author to pretend to be an agnostic about metamorphic facies. He believes that a broad facies scheme of the type expounded by Miyashiro is of great value for the discussion of metamorphism. He hopes that the student, after using this book in the course of his studies, will proceed to more advanced works in which a facies scheme forms the foundation for discussion.

14

Isotope geology of metamorphic rocks

UNSTABLE ISOTOPES

Radioactive isotopes with long decay half lives have been the subject of numerous investigations in igneous, sedimentary and metamorphic rocks. The aim of these studies is age determination, which is a field of Earth science outside the scope of this book. Much age determination work has been done on metamorphic rocks, especially those of Precambrian age. An understanding of the processes involved in metamorphism has been found to be essential for correct interpretation of the isotopic abundances in minerals and rocks in order to determine the age of metamorphism. Such studies therefore provide valuable additional information on processes and conditions of contact and regional metamorphism, although this is not their primary aim.

The unstable isotopes of four elements extensively used for age-dating are potassium 40 (^{40}K), rubidium 87 (^{87}Rb), thorium 232 (^{232}Th), uranium 235 (^{235}U) and uranium 238 (^{238}U). Potassium is the most abundant element in the list although ^{40}K makes up only a very small proportion of natural potassium. Rubidium is also present in many rocks as a trace element whose concentration is usually about ten parts per million. Uranium and thorium are rarer but tend to be concentrated in certain accessory minerals (e.g. zircon, apatite, orthite).

The nature of the **daughter isotopes**, which are the end-products of the radioactive decay of these elements, is crucial for the interpretation of concentrations of parent and daughter isotopes in metamorphic rocks. Figure 14.1 gives the final daughter isotopes for the parent isotopes listed above, and also gives the **half lives** for each decay sequence. In the cases of the uranium and thorium isotopes, this shows only the beginning and end isotopes at either end of complex decay sequences.

Argon, the element formed by the decay of ^{40}K, is an inert gas which does

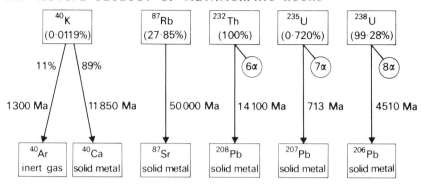

Figure 14.1   Radioactive decay of the principal unstable isotopes used to date metamorphic rocks. The upper boxes give the chemical symbols of the unstable parent isotopes and the percentage of atoms of the naturally occurring element which are of each unstable isotope. The lower boxes show the final stable daughter isotopes, and their physical nature in a pure state. The half lives for each type of decay are given alongside the arrow representing the decay. Circles show the number of α-particles emitted during the decay of uranium and thorium isotopes. For these isotopes the diagram summarises long, complex decay chains.

not combine chemically with minerals. At low temperatures it remains trapped in the crystal structures of minerals, but it is fairly readily driven out. Strontium is a metallic element, which can be incorporated into the crystal structure of certain minerals (such as plagioclase feldspars) as the divalent ion Sr^{+2}. The very small amounts of ^{87}Sr formed by the radioactive decay of ^{87}Rb can also be accommodated in the structures of minerals which do not normally contain strontium in defects in their crystal structure. ^{87}Sr is not as easily driven out of the rocks by heating as is ^{40}Ar. The lead isotopes produced by the decay of thorium and uranium are also strongly bound into the structure of minerals. Since zircon is extraordinarily resistant to solution in magmatic or metamorphic fluids, zircons often retain the ages of formation of rocks before metamorphism or even partial melting. They may even give the age of crystallisation of a parent igneous rock which was weathered to form sediment and subsequently metamorphosed.

It should be emphasised that radiometric measurements of all kinds on metamorphic rocks are more reliable if the history of crystallisation of the rocks has been carefully established by petrographic description and structural analysis. Many laboratories will not accept rocks for age determination unless this has been done. Discoveries about the conditions of metamorphism derived from the study of unstable isotopes are a byproduct of age determination. The complexity and cost of the apparatus and the specialist skill needed to operate it are such that this is likely to be the case for the foreseeable future.

THE POTASSIUM–ARGON METHOD

Potassium–argon age determinations measure the time since the rock or mineral analysed began to retain ^{40}Ar produced by the decay of ^{40}K. Different rocks and minerals have different abilities to retain ^{40}Ar. For example, potassium-rich glass and potash feldspar lose it rather easily, while hornblende tends to retain it.

^{40}Ar is lost from all solid materials by diffusion and therefore in all cases its loss is prompted by heating. The ability of a mineral to retain ^{40}Ar is

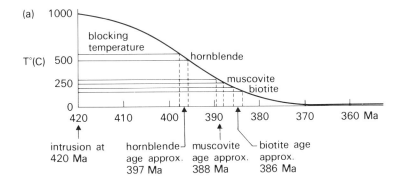

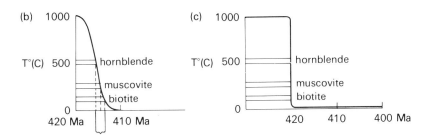

Figure 14.2 The relationship between potassium-argon dates on separated minerals from the same rock and the rate of cooling of the rock. (a) This rock has cooled slowly. Hornblende, muscovite and biotite give different ages of 386, 388 and 397 Ma respectively, because of their different blocking temperatures. (b) This rock has cooled during an interval of 10 Ma. Although in theory the hornblende, muscovite and biotite in this rock should give different ages, in practice the differences are smaller than the experimental uncertainty in potassium-argon age dating. The average value of the dates would be taken as the date of the metamorphic event. (c) This rock has cooled very rapidly over an interval of a few thousand years. Any differences in the mineral ages should be due to experimental uncertainty. Such rapid cooling is likely to be found in a shallow contact aureole, in which case the dates should be the same as the dates obtained from the igneous rocks of the associated intrusion. From Wilson (1971).

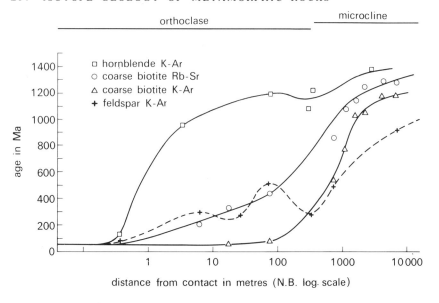

Figure 14.3 Variation of potassium-argon ages of separated minerals from the contact aureole of the Eldora Stock, Colorado, USA, plotted against distance from the contact in metres on a logarithmic scale. The extent of metamorphic zones characterised by contact metamorphic orthoclase and muscovite is shown by bars at the top of the diagram. From Hart (1964).

therefore conveniently expressed as a **blocking temperature** above which diffusion is sufficiently rapid for the argon to be driven off. Because the blocking temperatures of different minerals vary, the potassium–argon method can give an indication of the rate of cooling of a metamorphic rock (Fig. 14.2). Minerals with high blocking temperatures will give greater ages than those with low blocking temperatures if the rate of cooling is slow.

Complications arise if a rock has been reheated after initially cooling from a high temperature. If the reheating comes into the blocking temperature range there may be partial loss of argon, giving unreal 'ages' for the rocks or minerals, intermediate between the age of initial cooling and the age of reheating. Because of this interference with age determination results, studies have been made of potassium–argon ages in regional metamorphic rocks overprinted by contact metamorphic aureoles.

Hart (1964) studied the ages of minerals in a contact aureole in Colorado, USA, where Precambrian gneisses are intruded by a Tertiary quartz monzonite stock (Fig. 14.3). The differences in the potassium–argon ages obtained from biotite (1200 Ma) and hornblende (1400 Ma) in the schists 10 000 m from the contact probably reflect the slow cooling after regional metamorphism discussed above. The age of intrusion of the stock, determined from potassium–argon determinations on the minerals in the quartz monzonite, is 54 Ma. Using the blocking temperatures shown in Figure 14.2 it can be seen that the maximum temperature attained during reheating was about 500 °C at a distance of 3 m from the intrusion, and about 200 °C at 3000 m. The only isograd recognisable in the contact aureole, which is mainly overprinted on granitic gneisses, is marked by a change from microcline to orthoclase as the potash feldspar in the gneisses. This change takes place over a zone about 30 m wide at 325 m from the contact. The potassium–argon ages of the biotite crystals are altered much further from the contact than any mineral or textural change which can be recognised either in hand-specimen or with the petrological microscope.

Recently, a different technique for potassium–argon dating has been developed which gives more detailed information about argon loss from minerals. This is the ^{39}Ar–^{40}Ar spectrum method. The rock or mineral sample to be analysed is first irradiated with neutrons, converting a known proportion of the abundant stable isotope of potassium, ^{39}K, into ^{39}Ar. This isotope of argon does not occur naturally. The proportion of ^{39}Ar in the sample can be measured at the same time as the ^{40}Ar and the naturally occurring stable isotopes ^{38}Ar and ^{36}Ar which are present because of unavoidable contamination of the sample by atmospheric argon. The age and extent of atmospheric contamination can thus be determined, using the ^{39}Ar proportion to measure the ^{40}K contents, from the known ratio of ^{39}K to ^{40}K.

The special feature of the ^{39}Ar–^{40}Ar spectrum method is that the sample is heated in stages up to the temperature at which all the argon is driven out, and the proportions of the four isotopes of argon are separately determined at each stage. The heating is done in a near-vacuum. The age corresponding to each stage of heating can then be determined separately, the resulting curves being called **age spectra** (Fig. 14.4). The argon driven off at low temperatures shows a large amount of contamination by atmospheric argon, but in samples which have simply cooled once from well above their blocking temperatures, a plateau of constant age is reached in the high-temperature argon fractions. This plateau gives an age equal to the age of the sample determined from the total argon isotope ratios. Samples which have been reheated display more complex age spectra (Fig. 14.4). The interpretation of these is controversial. Spectra from meteorites and Moon rocks, which underwent reheating in a vacuum similar to the conditions of heating in the laboratory, can be interpreted to determine the ages of primary cooling and subsequent reheating and to estimate the maximum temperature attained during the reheating. Although the age spectra of reheated terrestrial metamorphic rocks are also disturbed when compared with the spectra of once-cooled samples, they are not so readily interpreted because their mechanism of argon loss has not been diffusion into a vacuum. The original cooling age can usually be determined from a plateau in the spectrum, but the age and temperature of reheating cannot be determined. There seems to be a good chance that the efforts of isotope

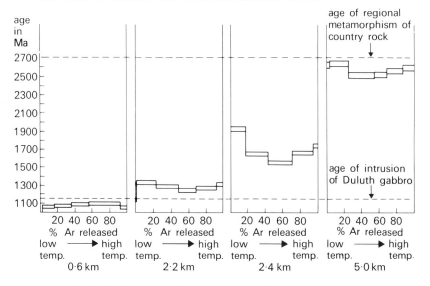

Figure 14.4 Variation in argon release spectra of biotites from the contact aureole of the Duluth Gabbro, Minnesota, USA. Distances of the samples from the contact are given below the spectra. Spectrum (a) is normal, giving the age of intrusion of the gabbro, spectrum (d) is normal, giving the age of regional metamorphism of the country rocks. Spectra (b) and (c) are disturbed. From Dallmeyer (1975) using data from Hanson *et al.* (1975).

geochemists to understand this phenomenon may lead to an advance in understanding of the role of diffusion in metamorphism.

THE RUBIDIUM–STRONTIUM METHOD

Because the daughter isotope of ^{87}Rb is ^{87}Sr, an isotope which is retained in the crystal structure of silicate minerals, the interpretation of ^{87}Rb and ^{87}Sr abundances in rocks and minerals in terms of their thermal histories is better understood than the abundances of the isotopes of argon. The rubidium–strontium 'clock' is less readily disturbed by reheating after primary cooling than the potassium–argon clock. Rubidium–strontium dates on minerals give the time which has elapsed since rubidium and strontium ceased to be able to diffuse freely throughout the rock sample and were differently concentrated into different minerals in the rock. This is because the chemical properties of rubidium and strontium are very different. A considerable proportion of ^{87}Sr was already present when the oldest rocks crystallised, so it is necessary to study the increase in the proportion of ^{87}Sr and the reduction in the proportion of ^{87}Rb in the sample. This is usually done by preparing **isochron diagrams**, plotting the ratio of ^{87}Sr to ^{86}Sr (a stable

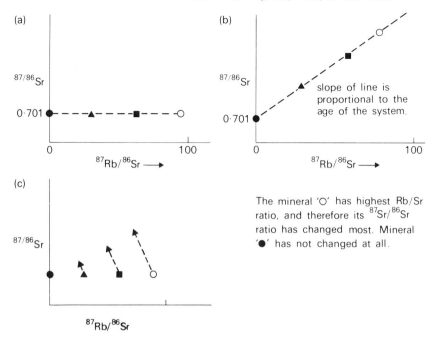

Figure 14.5 Explanation of isochron diagrams. $^{87}Sr/^{86}Sr$ ratios are plotted along the ordinate, $^{87}Rb/^{86}Sr$ ratios along the abscissa, for separated minerals from a rock. The whole rock ratios may also be plotted on isochron diagrams although they have not been plotted in this example. The closed circle represents a mineral containing no rubidium, the triangle, square and open circle minerals with increasing rubidium contents. (a) Ratios at the time of crystallisation. All four minerals have the same $^{87}Sr/^{86}Sr$ ratio, but vary in their $^{87}Rb/^{86}Sr$ ratio. (b) After several million years. The mineral with the highest Rb/Sr ratio (open circle) shows the largest change in its $^{87}Sr/^{86}Sr$ ratio. The mineral with no rubidium (closed circle) shows no change in the ratio. The points fall onto a straight line whose slope is proportional to the age of the system. (c) The dotted lines with the arrows show the change in the isotopic ratios of each mineral over the time interval between diagram (a) and diagram (b). From Wilson (1971).

isotope) vertically, and the ratio of ^{87}Rb to ^{86}Sr horizontally (Fig. 14.5). Because some minerals have a higher ratio of Rb to Sr than others, the final ratios of ^{87}Sr to ^{86}Sr will be higher in the rubidium-rich minerals (Fig. 14.5).

It is possible to plot isochron diagrams for individual rocks using several different minerals with different ratios of rubidium to strontium. The isotope ratios should define a straight line on the diagram whose slope gives a measure of the age. Such a line is called a **mineral isochron**. It is also possible to plot isochrons based upon several whole rock samples with different rubidium–strontium ratios. These are called **whole rock isochrons**.

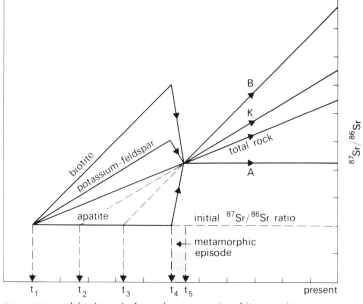

Figure 14.6 Isochron diagram for the minerals and whole rock in a granitic rock which has undergone metamorphism since its primary crystallisation. During the metamorphism from time t_4 to time t_5, the strontium isotope ratios became the same in all the minerals of the rock. The time interval t_4–t_5 is exaggerated relative to the rest of the history of the rock, compared with actual examples of this process. This has been done to emphasise that the homogenisation of the isotope ratios is not an instantaneous process. It is assumed that the Rb/Sr ratios of the minerals remained unchanged and the system was closed to strontium during metamorphism (i.e. there was no metasomatism). The dates calculated from the different minerals using an assumed initial $^{87}Sr/^{86}Sr$ ratio are discordant (e.g. t_2 and t_3) and have no geological significance. The only geologically significant date derivable from the data in the diagram is t_5, the time elapsed since the end of metamorphism. From Faure and Powell (1972).

Whereas a mineral isochron gives the time which has elapsed since rubidium and strontium were able to diffuse through a hand-specimen of rock, a whole rock isochron gives the time since rubidium and strontium could be exchanged over hundreds of metres or more. For a suite of rocks with a simple cooling history, such as a set of granites from one intrusion, mineral and whole rock isochrons give the same age. But in regional metamorphic rocks a whole rock isochron may give the age of primary cooling of the rocks, while mineral isochrons on individual samples give the age of subsequent recrystallisation. Metamorphism rarely makes the ratio of rubidium to strontium the same through large volumes of rock, but frequently does so for the minerals within individual rock samples (Fig. 14.6). This suggests that large-scale metasomatism involving rubidium and strontium is rare, a conclusion of considerable petrogenetic interest.

Rubidium−strontium age dating has been of particular value in the study of metamorphic rocks which have undergone several cycles of metamorphism. Whereas the textures and structures of the rocks are open to several interpretations, rubidium−strontium isochron diagrams can sometimes provide unambiguous evidence for more than one episode of metamorphism. However, from time to time there have been quite acrimonious disputes among geochronologists about the interpretation of age-dating results from metamorphic rocks. This happens when there is a scatter of ages, especially if the results of different age-dating techniques are different. In part, this arises from a lack of knowledge about diffusion processes in metamorphism.

STABLE ISOTOPES

Studies of the ratios of different isotopes of oxygen and hydrogen in metamorphic rocks have permitted estimates of the temperature of metamorphism in favourable cases, and have also provided most valuable information about the origin of the water in hydrous metamorphic minerals such as micas and amphiboles. Unlike the isotopes discussed in the previous part of this chapter, the naturally occurring isotopes of oxygen and hydrogen are stable and do not decay to other nuclides; nor are they created by the decay of other unstable nuclides.

The ratios of the isotopes differ in samples of oxygen and hydrogen from different sources. This variation in the isotopic composition is due to fractionation of the isotopes during the continuous chemical evolution of the Earth. In the case of hydrogen and oxygen in rain water, such fractionation occurs constantly during the hydrological cycle. For water locked in the crystal structure of metamorphic rocks, there was fractionation of the different isotopes at the time of metamorphism. The proportions of the different stable isotopes of hydrogen and oxygen in sea water are as follows: ^1H 99·985%, D (^2H) 0·015%; ^{16}O 99·759%, ^{17}O 0·037%, ^{18}O 0·204%.

The variations in isotopic ratios are small. Two ratios are commonly measured − that of deuterium to hydrogen, and that of ^{18}O to ^{16}O. These ratios are compared with a standard ratio. Because the isotopic compositions of both hydrogen and oxygen are studied, the standard used is water. The oceans provide the reservoir of water for the hydrological cycle and therefore the average isotopic composition of ocean water is selected. This standard is called **SMOW** (Standard Mean Ocean Water).

The relative concentration D/H or $^{18}O/^{16}O$ in a particular sample is expressed by the factor δ (delta), defined as follows:

for D

$$R_{sample} = D/H \text{ in sample}$$
$$R_{SMOW} = D/H \text{ in SMOW}$$
$$\delta D = (R_{sample}/R_{SMOW} - 1).\ 1000 \qquad (1)$$

for ^{18}O

$$R_{sample} = {}^{18}O/{}^{16}O \text{ in sample}$$
$$R_{SMOW} = {}^{18}O/{}^{16}O \text{ in SMOW}$$
$$\delta^{18}O = (R_{sample}/R_{SMOW} - 1).\ 1000 \qquad (2)$$

Positive values of δ show that the sample is enriched in deuterium or ^{18}O compared with sea water, negative values that it is depleted in these isotopes.

In rocks, the differences of $^{18}O/{}^{16}O$ in different minerals are expressed by the ratio α (alpha). For example, the ratios might be determined in coexisting muscovite and quartz in a schist. In this case:

$$R_{muscovite} = {}^{18}O/{}^{16}O \text{ in muscovite}$$
$$R_{quartz} = {}^{18}O/{}^{16}O \text{ in quartz}$$
$$\alpha_{muscovite-quartz} = R_{muscovite}/R_{quartz} \qquad (3)$$

If the concentration of ^{18}O in the minerals are quoted as δ values, α may be found from the relation:

$$\alpha_{muscovite-quartz} \simeq \exp\ \{(\delta_{muscovite} - \delta_{quartz})/1000\} \qquad (4)$$

OXYGEN ISOTOPE GEOTHERMOMETRY

At temperatures greater than about 500 °C, the fractionation factor α for pairs of minerals coexisting in equilibrium is directly related to the temperature of fractionation of the oxygen isotopes. In most of the metamorphic rocks studied, this temperature appears to be that at which the equilibrium metamorphic assemblage crystallised (i.e. the temperature of metamorphism discussed in this book). Temperature determinations by this method are not influenced by pressure or composition of volatile phases participating in the metamorphic reactions. This is a valuable advantage over mineralogical geothermometry as described in Chapter 13. The experimental techniques involved are difficult and comparatively few determinations have yet been done. There are problems in ensuring that the determinations are carried out on equilibrium mineral assemblages. Retrogressive changes in the oxygen isotope composition of metamorphic minerals can occur, without corresponding changes being visible under the petrological microscope.

The most comprehensive study of oxygen isotope ratios of regional metamorphic rocks is that of Garlick and Epstein (1967). They determined the δ values for ^{18}O in coexisting minerals of rocks from several of the best-known progressive regional metamorphic sequences in the USA. The majority of the rocks they studied were pelitic in composition and the sequence of isograds in the areas concerned are of the Barrow type, like those of Sulitjelma described in this book. They also analysed the minerals of quartz and pegmatite veins found among the regional metamorphic rocks.

They discovered that the order of increasing $\delta^{18}O$ in the minerals of a metamorphic rock was almost always the same. This enabled them to construct a sequence of minerals according to their tendency to concentrate ^{18}O as follows:

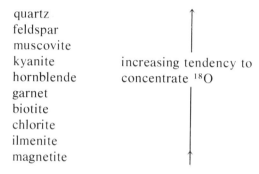

Where the samples they studied come from locations only a few hundred metres apart, they found that the $\delta^{18}O$ values of the minerals were similar, implying that the oxygen isotopes had equilibrated in their concentrations through rather large volumes of rock. This conclusion was supported by the similarity of $\delta^{18}O$ in veins and in their host rocks, indicating that the veins formed by sweating out of fluid with dissolved mineral constituents during metamorphism. The exceptions to this widespread homogenisation of oxygen isotope ratios were a schist from close to a marble layer and a metamorphosed dyke of amphibolite. In both cases there are reasonable geological grounds to explain the difference. In the specimen close to the marble, the minerals presumably fractionated oxygen not from H_2O but from an H_2O-CO_2 mixture. In the amphibolite the isotope ratios might still be influenced by the isotope composition of the primary basaltic magma.

The temperature determinations which Garlick and Epstein made were based upon experimental work on the pair of minerals giving the highest α value, which from the list above can be seen to be quartz and magnetite. Figure 14.7 shows a plot of α values for various mineral pairs against temperature for rocks of different metamorphic grades, defined in terms of the traditional Barrow index minerals. This diagram could be used with

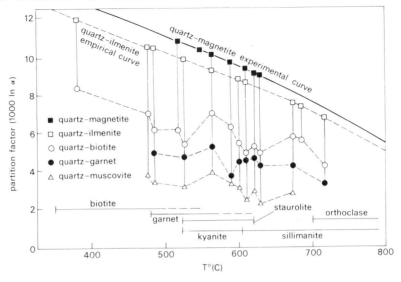

Figure 14.7 Plot of oxygen fractionation factor against temperature for various mineral pairs with increasing metamorphic grade. The indicated temperatures are based upon experimental values for quartz-magnetite fractionation (solid line) and empirical values for quartz-ilmenite fractionation (dashed line). From Garlick and Epstein (1967).

values of $\alpha_{\text{ilmenite-quartz}}$ for temperature determination. Other mineral pairs could be used, but as α diminishes the precision of the determination would diminish also.

THE ORIGIN OF WATER IN METAMORPHIC ROCKS

Meteoric water, that is rain water or snow, has a limited distinctive range of hydrogen and oxygen isotope ratios, related to the mean annual temperature and therefore the latitude of the point where it falls. On a diagram plotting δD against $\delta^{18}O$, these meteoric waters define a straight line (Fig. 14.8). This line is called the **meteoric water line**. Primary magmatic water, which has come from the source region of magmas in the upper mantle, has a comparatively small range of isotopic compositions, also shown on Figure 14.8. This composition of primary magmatic water is hard to determine experimentally because exchange of oxygen and hydrogen isotopes with circulating ground water after the magma has crystallised is very common, even in rocks which show no petrographic sign of alteration. The water from metamorphic rocks shows a wider range of isotopic compositions. The well-defined isotopic compositions of meteoric waters and primary magmatic waters have proved to be very valuable in the study of contact metamorphic rocks (Taylor 1974).

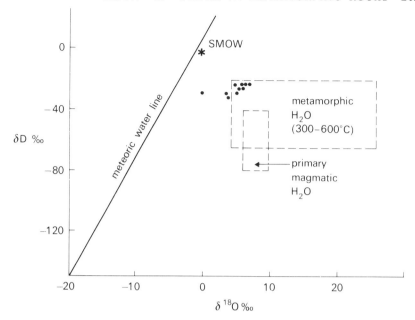

Points are possible modern metamorphic waters from California

Figure 14.8 Oxygen and hydrogen isotopic variation in meteoric waters (line) primary magmatic waters and metamorphic waters (Fields). From Taylor (1974).

The values of $\delta^{18}O$ have been more widely used than the values of δD, or the combination of the two (which gives the least ambiguous results). In the rocks inside the Skaergaard Intrusion, East Greenland, the lower parts show magmatic values of $\delta^{18}O$, while $\delta^{18}O$ values from plagioclases in the upper part of the intrusion are lower, nearer to meteoric values. The intrusion cuts across a major geological boundary between permeable Tertiary basalt flows and impermeable Precambrian gneisses, and this boundary corresponds in level to the change in the $\delta^{18}O$ values from plagioclase feldspars from within the intrusion. This observation may be explained by circulation of ground water through the basalts and the upper part of the intrusion, exchanging oxygen isotopes with the plagioclase feldspars during cooling but after the solidification of the intrusion (Fig. 14.9). This suggests that earlier, while the gabbro intrusion was still solidifying, ground water also circulated through the permeable basalts. It was probably unable to interact isotopically with the differentiating magma because of the 'armouring' of the intrusion by an impermeable skin of chilled magma. After solidification, fracturing during cooling permitted ground water to percolate through the upper part of the intrusion.

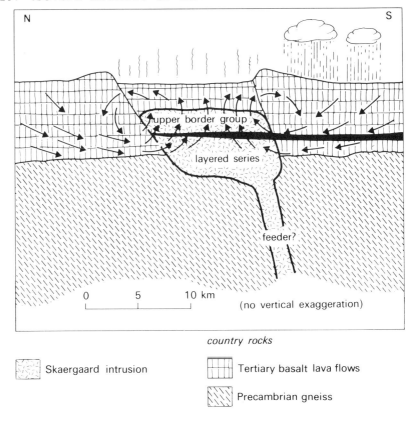

country rocks

▨ Skaergaard intrusion   ⊞ Tertiary basalt lava flows

                              ◪ Precambrian gneiss

Figure 14.9 Schematic cross-section through the Skaergaard Intrusion, East Greenland, showing the meteoric-hydrothermal water circulation in the Plateau Lavas above the gneiss basement. Only minor amounts of meteoric-hydrothermal water penetrated the gneiss basement along fractures. From Taylor (1974).

Transport of heat by circulating ground water, as illustrated by this example, is probably the most important mechanism for the loss of heat from cooling intrusions at high levels in the Earth's crust. This explains their narrow and comparatively irregular contact aureoles. The circulating ground water was relatively efficient at removing heat from the outer aureole, while there were probably steep temperature gradients in the narrow chilled margin and inner aureole which were less permeable to ground water or impermeable. Heat loss through this narrow zone would be by conduction.

Of the contact aureoles discussed in Chapter 5, only the aureole of the Beinn an Dubhaich granite may have formed at the level of ground-water circulation, from the evidence of the regional geology. This is supported by some data on oxygen isotope compositions (Taylor & Forester 1971). Comrie and Skiddaw almost certainly formed deeper in the crust, below the level at which ground water circulated. In both cases, the metamorphic reactions show that water fixed in the crystal structures of phyllosilicates was lost from the rocks during metamorphism. The mechanism by which this occurred is not known, but diffusion along grain boundaries during metamorphism is a possibility.

Oxygen and hydrogen isotope studies have also led to important conclusions about ocean-floor metamorphism (Ch. 11). Determination of the isotopic ratios of both hydrogen and oxygen from rocks of the Troodos Complex, Cyprus, shows that the water in the hydrous minerals is closer in isotopic composition to sea water than to primary magmatic water (Spooner *et al*. 1974). Sea-water isotopic compositions are found through the Pillow Lava and Sheeted Dyke Complex, but change to magmatic values in the gabbros and cumulate ultrabasic layers. This obviously supports the view of the metamorphism of the Troodos Complex put forward in Chapter 11.

The application of hydrogen and oxygen isotope studies to metamorphic rocks is still in its preliminary stages. The conclusions already reached are most impressive and suggest that these techniques could revolutionise understanding of many metamorphic processes. It is worth emphasising that, like radiometric age determination, stable isotope studies are most meaningful when carried out on rocks whose petrology, mineralogy and geochemistry have been carefully studied, like those of Skaergaard and Troodos. The introduction of new geochemical techniques does not render petrographic study obsolete; it makes it more important.

15

Electron-probe microanalysis

Electron-probe microanalysis determines the chemical composition of a small volume at the surface of a polished mineral preparation. Under the best conditions the area above the volume analysed is a circle approximately $1\mu m$ in diameter. The apparatus used to perform the analysis is elaborate and expensive. Modern instruments have however become sufficiently reliable and simple in operation for the operator not to need specialist knowledge of physics and electronics. They are frequently operated by postgraduate geology students, and sometimes by undergraduates taking advanced courses in mineralogy.

The reason for including an outline of this comparatively advanced technique in this elementary book is that it provides a unique opportunity to supplement petrographic study under the microscope by chemical analysis, which can even be carried out on the same grains. Electron-probe microanalysis can be carried out on thin sections provided that they are not covered by a cover slip and have been polished. Figure 6.11 was drawn from such a thin section. The thin section to be analysed has to be coated with a layer of a conducting material, usually a very thin film of graphite. This does not interfere with the familiar techniques for examination in transmitted light although the higher relief which results from the absence of a cover slip may be confusing. Electron-probe microanalysers designed for use in Earth sciences usually have a petrological microscope built in, enabling the specimen for analysis to be examined in plane-polarised light and between crossed polars. However, it is not usually possible to rotate the specimen.

THE ELECTRON-PROBE MICROANALYSER

The electron-probe microanalyser focuses a very narrow beam of energetic electrons onto the surface of a specimen. The beam strikes the surface at the $1\ \mu m$ diameter spot mentioned earlier. The electrons excite the atoms of the specimen to emit X-rays. The frequencies and energies of the X-rays emitted are characteristic of the elements present in the specimen, and their intensities are approximately proportional to the concentration of each element.

The X-rays characteristic of each element have to be distinguished. This may be done in one of two ways. X-rays of different wavelengths may be separated using one or more spectrometers and the intensity of each characteristic wavelength measured by a suitable detector. This method is the one used in most instruments in routine use at the time of writing (1977). Alternatively, the X-rays may be detected without being separated into different frequencies, using a detector which measures the energy as well as the intensity of the X-rays. The distribution of energies may then be determined electronically and the elements present in the sample recognised by concentrations of energies at particular values. This has become possible recently because of improvements in X-ray detectors and in the electronic equipment needed to separate the different energies. An instrument of this sort is simpler mechanically, because it lacks the spectrometers, but requires more advanced electronics. It is likely to be more widely used in future. Figure 15.1 is a schematic diagram of the two types of instrument. The apparatus producing and focusing the electron beam is the same in each and is very similar to an electron microscope.

At the appropriate wavelength or energy peak, the intensity of the X-rays produced is roughly proportional to the concentration of an element in the sample. Analyses are performed by comparing the intensity of X-rays produced from the sample with those produced by a standard under the same conditions. The standard is a substance in which the concentration of the elements being analysed is known. It may be pure metal, a mineral analysed by other methods or analysed synthetic silicate glass. For accurate analysis, a number of corrections need to be applied to the X-ray intensity measurements made by the instrument (Long 1967) and on apparatus without a built-in computer the results are usually sent away to be corrected. The rough proportionality of X-ray intensity and concentration is usually good enough for a mineral to be identified without waiting for the results from the computer, a fact which seems to escape some students! The energy dispersive apparatus employs its own computer and presents the results already corrected.

If carefully used, the electron-probe microanalyser can perform mineral analyses of an accuracy comparable to those made by chemical analysis or other physical analytical techniques such as X-ray fluorescence. It can analyse several spots in large mineral grains, revealing and measuring zoning. It can analyse small grains, down to less than 5 μm across.

THE STUDY OF METAMORPHIC REACTIONS

The electron-probe microanalyser can provide a check on the assumption

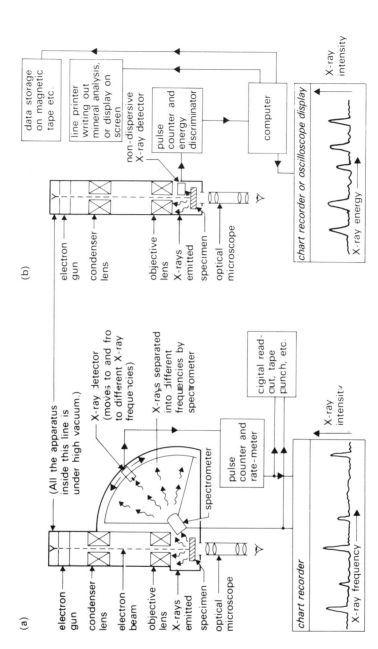

(a)

(b)

electron
gun

condenser
lens

electron
beam

objective
lens

X-rays
emitted

specimen

optical
microscope

(All the apparatus
inside this line is
under high vacuum.)

X-ray detector
(moves to and fro
to different X-ray
frequencies.)

X-rays separated
into different
frequencies by
spectrometer

spectrometer

pulse
counter and
rate-meter

digital read-
out, tape
punch, etc.

chart recorder

X-ray
intensity

X-ray frequency

electron
gun

condenser
lens

objective
lens

X-rays
emitted

specimen

optical
microscope

data storage
on magnetic
tape etc.

line printer
writing out
mineral analysis,
or display on
screen

non-dispersive
X-ray detector

pulse
counter and
energy
discriminator

computer

chart recorder or oscilloscope display

X-ray
intensity

X-ray energy

Figure 15.1 The electron-probe microanalyser. (a) Wavelength dispersive system. (b) Energy dispersive system. After Long (1967).

that a metamorphic rock represents an equilibrium mineral assemblage. It may be used to check whether any of the minerals in the rock, transparent or opaque, are zoned. The minerals can be identified, including the opaques. The compositions of minerals with a range of atomic substitutions, such as amphiboles, epidotes and micas, can be determined and the distribution of elements between them determined. The use of this information to determine the temperature and pressure of metamorphism was discussed in Chapter 13.

The result of such studies generally confirm the assumption of equilibrium in most metamorphic rocks which meet the criteria given in Chapter 4. Rocks have of course been studied in which the microprobe indicates that metamorphic reaction is incomplete (for example the two eclogites from kimberlite discussed in Ch. 12) but this conclusion usually confirms the results of textural studies or phase rule considerations. In most metamorphic rocks, the minerals are unzoned, even in cases where their textures indicate a complex history of growth during the deformation of the rocks (Ch. 9).

There is one notable exception to this rule. Almandine-rich garnets of medium-grade metamorphic rocks are frequently zoned (Harte & Henley 1966, Atherton 1968). The zoning patterns show broad similarities, but some minor differences, in the different cases described. In some cases there are sharp breaks in the zoning pattern, perhaps indicating a pause in garnet growth followed by growth under different conditions. Many garnets have a regular increase of manganese content into the core, matched by a corresponding decrease in iron. This was noticed early in the study of garnet zoning (Harte & Henley 1966), but there is no general agreement on its cause.

One view is that the zoning is due to fractionation of manganese into the garnet during its growth (Hollister 1966, Atherton 1968). On this view, manganese or any other element, once it has been incorporated into the garnet, is removed from the crystallising rock system. In other words, there is no diffusion at all of material into and out of the garnet crystals. Manganese tends to be very strongly concentrated in the garnet rather than the other minerals in the rock. The combination of these two processes explains the form of the manganese zoning quite well. Atherton's model also permits an estimate of the distance the manganese has diffused through the rock to reach the garnets. In some cases this is as much as 2–3 mm.

An alternative theory is that the zoning is caused by diffusion of elements into and out of the garnet crystals as they approached, but did not achieve equilibrium with, the surrounding mineral assemblage (Anderson & Buckley 1973). This view is favoured by the fact that although the garnets have rather irregular boundaries as they selectively replace the grain boundaries of the surrounding minerals (Fig. 8.6), the boundaries between the layers of different composition in them are not irregular. The issue is likely to be

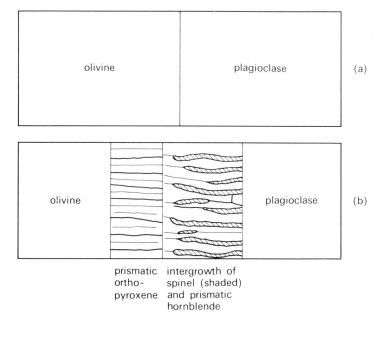

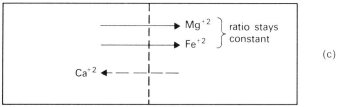

Figure 15.2 Development of coronas in troctolite from Sulitjelma. (a) Original igneous grain boundary between olivine and plagioclase feldspar. (b) Coronas form by partial reaction during slow cooling. (c) Diffusion of ions across the original boundary needed to explain the corona compositions.

resolved by careful textural studies of the garnets combined with micro-probe analyses. It is particularly important to study the form of zoning in three dimensions, using serial sections through large garnets. Work of this kind is now in progress.

THE STUDY OF CORONAS

Coronas are reaction rims found in certain igneous rocks. They have a uniform thickness and are usually formed of prismatic crystals whose long axes are perpendicular to the rims themselves. Their textural features distinguish them clearly from magmatic reaction rims and leave little doubt that they formed after solidification of the igneous rock, in the solid state.

The author has studied coronas occurring round olivine crystals in a troctolite from the marginal facies of the Sulitjelma Gabbro, Norway (Mason 1967). They are the product of metamorphic reaction between olivine and the surrounding plagioclase feldspar. By analysing each rim using the electron-probe microanalyser it was possible to show that the coronas had formed by diffusion of Mg and Fe^{+2} out of the olivine into the plagioclase, accompanied by a little diffusion of Ca in the opposite direction (Fig. 15.2). Not all coronas are products of simple two-way diffusion of this kind. But these coronas show how in suitable cases the use of the electron-probe microanalyser can permit quite a precise description of a metamorphic reaction.

The instrument has also been used to provide many more mineral ana-lyses than was previously possible for delineating isograds and defining coexisting mineral assemblages. Its potential as a tool for research into metamorphic processes is very great, so that it is likely to be used more extensively and, the author hopes, rather more imaginatively in the next few years.

16

Metamorphic rocks and the evolution of the Earth

In the previous chapters, examples have been outlined in which the study of metamorphic rocks has permitted the reconstruction of temperature and pressure at depth in the Earth in past geological times. This has been done for rocks from both oceanic and continental areas, and from ancient and more modern orogenic belts. In Chapter 12, it was shown that this kind of study is beginning to be extended to the Earth's mantle. In Chapter 6, it was demonstrated that the study of metamorphic rocks may also illuminate mechanical processes operating in the Earth. All these studies provide important information to be considered as part of a wider study of the Earth, especially of processes operating in the deeper parts of the crust and the upper mantle.

The outlines of global tectonic theory, usually referred to as **plate tectonics**, should be familiar to the reader, and it is not proposed to repeat them here. Reviews are to be found in Gass *et al*. 1971, Wyllie 1971, Oxburgh 1974, and numerous other works. Studies of metamorphic rocks have illuminated the theory, especially in regard to processes operating beneath mid-ocean ridges and island arcs.

The rocks formed by metamorphism at mid-ocean ridges have been described in Chapter 11. In Chapter 11 and Chapter 14 it was shown that metamorphism and perhaps metasomatism of the rocks of the oceanic crust is associated with sea water circulating through the rocks. This circulation occurs right down to the base of the oceanic crust. Heat is probably lost from mid-ocean ridge regions more by convection of this circulating sea water than by conduction through the solid rocks. This accounts for the variation in heat flow readings obtained from mid-ocean ridges, but also makes it difficult to arrive at average values for the rate of heat loss. The metamorphic assemblages of mid-ocean ridges indicate that at the base of the oceanic crust the temperature is about 500 °C. A geothermal curve is plotted on Figure 16.1 based upon the metamorphic assemblages of the

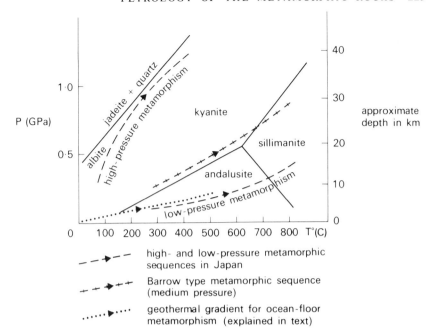

Figure 16.1 Different temperature-pressure gradients in metamorphism. From Miyashiro (1973) with ocean-floor geothermal gradient added.

Troodos Complex, Cyprus, but with a correction for the thickness of oceanic crust because the Troodos crust seems to be unusually thin. The overall gradient is shown, with the disturbing effects of the convection cells eliminated. The steep geothermal gradient (temperature rising rapidly with depth) supports the conclusion that the average rate of heat flow through mid-ocean ridges is higher than for the oceans as a whole.

Study of metamorphic rocks similarly permits a discussion of the temperature gradients beneath island arcs. This was appreciated before the development of geotectonic theory. In the circum-Pacific island arcs, parallel belts of metamorphic rocks of approximately the same age show contrasting types of metamorphism. This has been most convincingly demonstrated in Japan (Fig. 16.2) by Miyashiro and his co-workers. The belts occur in pairs: the member on the oceanic side developing blueschists and eclogites at high grade from rocks of basic igneous composition, the belt on the continental side developing andalusite and sillimanite at high grade from rocks of pelitic composition. In both belts there are considerable variations in metamorphic grade. In the oceanic side belt the geothermal gradients determined from study of the metamorphic assemblages are low (Fig. 16.1), so that at any given temperature the pressure is relatively high. These are therefore called high pressure metamorphic belts. In the con-

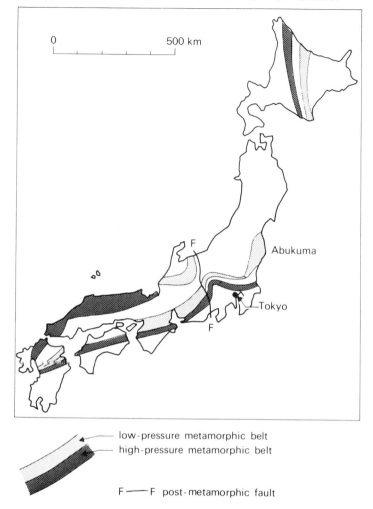

Figure 16.2 Paired metamorphic belts of Japan. From Miyashiro (1973).

tinental side belt the geothermal gradients are high (Fig. 16.1) so at any given temperature the pressure is relatively low. These are therefore called low pressure belts.

The distribution of temperature and pressure gradients is well explained by the dragging down of ocean floor material beneath the island arc with an unusually low geothermal gradient (Fig. 16.3). The pressure in the descending oceanic crust rises rapidly as it is dragged down into the Earth, while the temperature rises more slowly because of the slow conduction of heat into the descending slab. The relatively high geothermal gradient in the island arc itself is caused by unusually high heat flow. This is associated with

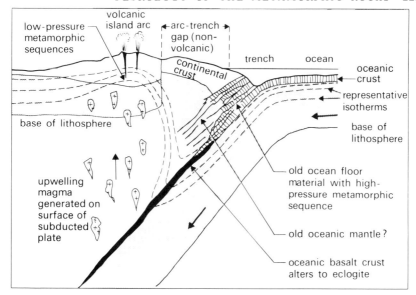

Figure 16.3 Schematic model of lithosphere beneath an island arc. The large arrows show the motion of the oceanic lithosphere relative to the continental lithosphere on the left. The close spaced isotherms in continental crust beneath the volcanic part of the island arc indicate a high geothermal gradient. From Miyashiro (1973) with sketch isotherms added.

volcanic activity, both as surface volcanoes and as intrusions at depth. The magmas are acid, intermediate and basic in composition and studies of their geochemical variation indicate that some are generated close to the upper surface of the descending plate. The immediate cause of the high heat flow is probably the uprise of these magmas into the crust. The ultimate source of the heat is at the surface of the descending plate. Oxburgh and Turcotte (1971) argue that it is produced by frictional heating, but as no metamorphic rocks are known which might be derived from this depth and tectonic setting, their ideas cannot be tested by metamorphic petrology.

The model of the crust and upper mantle beneath an island arc shown in Figure 16.3 is one of the most satisfactory which has yet been developed in explaining a multitude of features — geophysical, volcanic and metamorphic — of island arcs. It illustrates admirably the virtue of plate tectonics in integrating the results of several branches of Earth science, including metamorphic petrology.

Attempts to arrive at a similar integration for mountain belts in continental areas, such as the Alps or the Caledonides, have been less satisfactory. High-pressure metamorphic sequences grading to blueschists and eclogites occur in the Alpine orogenic belt, as described in Chapter 10, but they are associated not with low pressure sequences with andalusite and

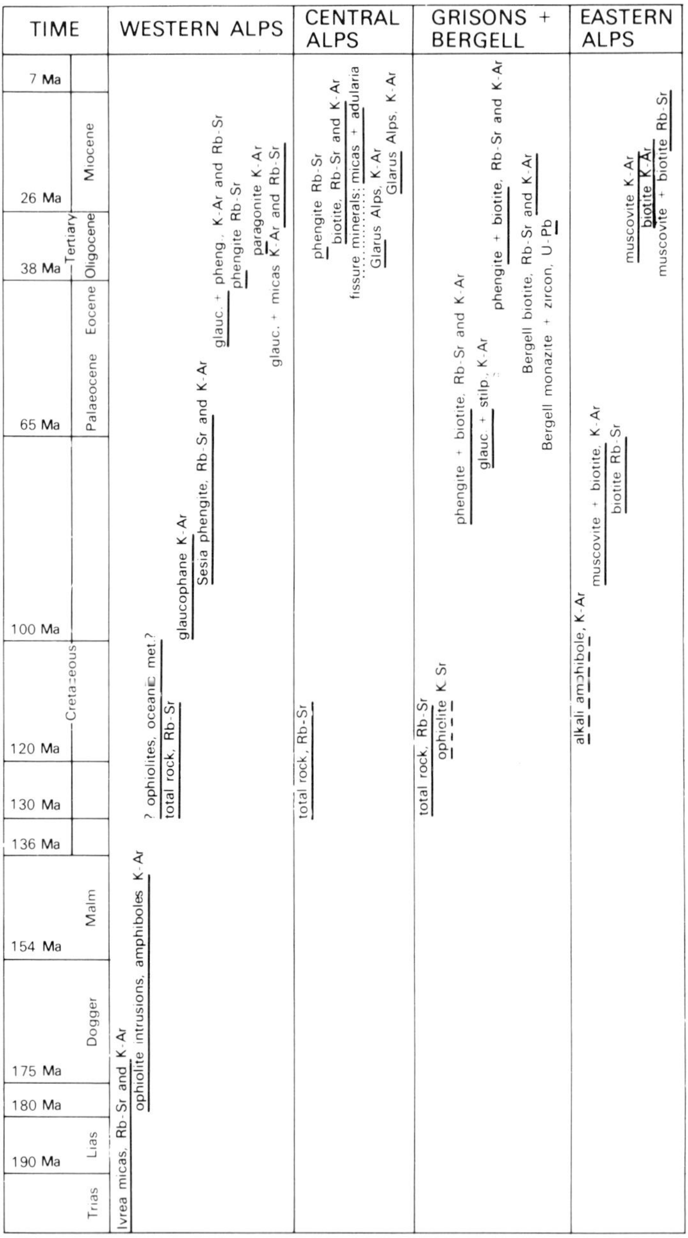

Figure 16.4 Radiometric dates of crystallisation of metamorphic rocks from different parts of the Alps. The vertical bars show the spread of dates from the minerals or rocks described on their left. Dashed bars represent speculative dates. From Frey *et al.* (1974).

sillimanite at high grade, but with Barrow-type sequences with kyanite and sillimanite. These may be described as medium-pressure sequences (Fig. 16.1). There is not a clear pattern of metamorphic belts. The metamorphosed parts of the Caledonide chain contain abundant medium-pressure metamorphic sequences, but blueschists are known only from two small localities. It has been suggested (Dewey 1969) that paired metamorphic belts are present, the medium-pressure metamorphic belt being equivalent to the low-pressure metamorphic belt of the Japanese islands, and the high-pressure belt being mostly buried under younger sedimentary rocks or cut out by faulting. Plate tectonic models for continental orogenic belts are becoming increasingly elaborate (Smith 1976) and the rather broad conclusions of metamorphic studies are less applicable to such models than to the simple island arc model. More detailed studies of metamorphism can be applied, and some pioneering efforts have been made in this direction. A model for temperature distribution in the eastern Alps, east of the area discussed in Chapter 10, has been developed by Oxburgh and Turcotte (1974). It has the merit of explaining the observation that early blueschist assemblages become replaced by later amphibolite assemblages, implying an early low geothermal gradient, succeeded by a later high one. In both the Pennide Nappes of the central Alps (Ch. 10) and in Connemara, Ireland (Ch. 9) early high-pressure metamorphism seems to have been succeeded by later, lower-pressure metamorphism. Perhaps Oxburgh and Turcotte's model is of wider application than to the eastern Alps. The writer hopes that there will be progress in this field of plate tectonic theory in the future.

The time relationship between deformation and metamorphism is obviously important in this context. Figure 16.4 gives the times of metamorphism in the Alps, the orogenic belt probably best understood in this respect. The early blueschist metamorphism and later Barrow-type regional metamorphism are strikingly shown. There appears also to be evidence of an earlier phase of ocean-floor type metamorphism. This sequence of metamorphic events would be well explained by a model for orogenic deformation involving initial rifting and formation of ocean floor, followed by destruction of the ocean floor by subduction, and finally by continental collision. Such a model is shown in Figure 16.5, and was first proposed by Dewey and Bird (1970). A modification of their original proposal is that in the Alps it is unlikely that the rifting continued long enough to form a wide ocean, only a comparatively narrow rift.

It is likely that metamorphic recrystallisation of the crust and upper mantle has an important role in the crustal shortening in orogenic belts. This is a departure from pure plate tectonic theory, which assumes that all relative movement between slabs of the **lithosphere** is taken up on fracture surfaces and that the internal parts of the lithospheric plates are undeformed. The model for the evolution of an orogenic belt illustrated in Figure 16.5 is fairly 'pure' in this sense.

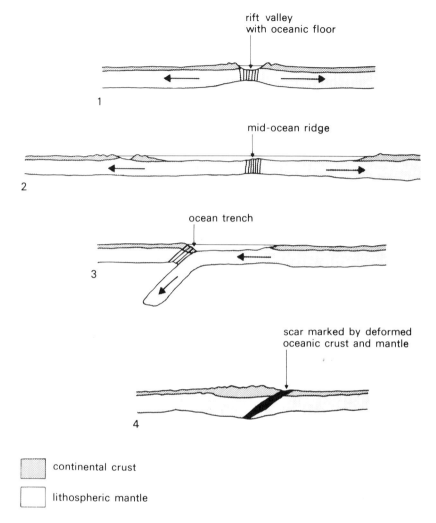

Figure 16.5 The development of a mountain chain by oceanic rifting followed by continental collision. 1 Initial rifting apart of continent. 2 Widening oceanic rift. This stage is perhaps represented by the radiometric dates in the range 130–100 Ma from ophiolitic rocks of the Western Alps (Fig. 16.4). 3 Oceanic rift closing by subduction. The mid-ocean ridge has disappeared down the subduction zone. The structure above the subduction zone is shown in more detail in Figure 16.3, and an alternative model for this stage and the next is proposed in Figure 16.6. This stage is perhaps represented by the glaucophane crystallisation dates from the Alps in the range 100–60 Ma (Fig. 16.4). 4 Relative movement of plates has ceased after continental collision. This stage is perhaps represented by the numerous radiometric dates from the Alps more recent than 38 Ma (Fig. 16.4). From Mason 1975).

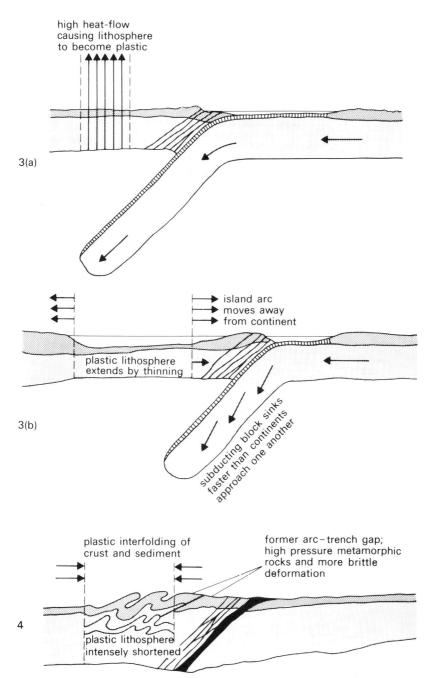

high heat-flow
causing lithosphere
to become plastic

3(a)

island arc
moves away
from continent

plastic lithosphere
extends by thinning

3(b)

subducting block sinks
faster than continents
approach one another

plastic interfolding of
crust and sediment

former arc–trench gap;
high pressure metamorphic
rocks and more brittle
deformation

4

plastic lithosphere
intensely shortened

Figure 16.6 Alternative model for stages 3 and 4 of Figure 16.5. 3(a) High heat flow associated with volcanism in the island arc causes a strip of lithosphere to become plastic. 3(b) Extension of plastic strip of continental lithosphere occurs by oceanward migration of the island arc. This is caused by the subducted lithospheric slab sinking faster than the continents approach. 4 Last stage of deformation after continental collision. The plastic strip of lithosphere, previously extended, is intensely shortened after continental collision has caused relative motion of plates to cease at the subduction zone.

However, in the interior parts of orogenic belts, such as the Pennide Nappes of the Alps, continental crustal material which had already been metamorphosed to gneiss before the onset of the orogeny, has been folded with the sediments overlying it into immense isoclinal sheets. The progressive onset of deformation in gneisses as orogenic belts are approached has been well documented in Precambrian shield areas. It appears that in orogenic belts a welt of crustal material as much as several hundred kilometres wide may become plastic during orogeny and may be intensely shortened during the final stages of continental collision. The plasticity is presumably the result of heating during metamorphism, perhaps aided by the introduction of H_2O from the subducted slab of lithosphere. Figure 16.6 shows a modification of the final stages in orogenic development from Figure 16.5 to allow for this crustal shortening. It has one important consequence. Many plate tectonic theories assume that the area of continental crust is increasing through geological time by addition of continental material at orogenic belts. This theory is called **continental accretion**. If there is extensive mobilisation and deformation of continental crust, as shown in Figure 16.6, it may be that the total area of continental crust has increased little during Phanerozoic time at least. The area added to the continental crust during accretion above subduction zones may be balanced by the area lost by crustal shortening elsewhere in the orogenic belt, although there will be an overall increase in the thickness of continental crust.

It has been argued that the widespread occurrence of pyroxene gneisses in the older parts of Precambrian shield areas may indicate that geothermal gradients were higher in early Precambrian times than they are today. The variation in geothermal gradients indicated by metamorphic assemblages in Phanerozoic rocks (Fig. 16.1) makes it difficult to judge whether this is the case. The gradients postulated for the early Precambrian crust can be matched in Phanerozoic crust, so the argument concerns a comparison of average gradients in Precambrian and Phanerozoic crust. The ancient Precambrian remnants of crust may be areas of Precambrian crust with a high geothermal gradient during metamorphism which have survived better than those where geothermal gradients were lower. Some support for this view can be found in the limited occurrence of blueschists in rocks older than the Cainozoic.

The value of metamorphic petrology in providing information about ancient geothermal gradients has brought it nearer to other branches of Earth science than it used to be. The importance of sound petrological study of specimens of metamorphic rock used for geochemical analysis and radiometric age-dating has also been emphasised. Metamorphic petrology is no longer the province of a few enthusiasts and should be part of the training of all Earth scientists.

Appendix

Table of molecular weights to use in calculations of rock and mineral compositions on composition–assemblage diagrams.

Al_2O_3	101·82	MnO_2	86·93
B	10·81	Mn_3O_4	228·79
B_2O_3	69·60	Na_2O	61·97
BaO	153·33	NiO	74·70
BeO	25·01	Nb_2O_5	265·78
C	12·01	P_2O_5	141·92
CO_2	44·00	PbO	223·18
CaO	56·07	Rb_2O	186·93
CeO_2	172·11	S	32·06
Ce_2O_3	328·22	SO_3	80·05
Cl	35·45	Sc_2O_3	137·89
CoO	74·93	SiO_2	60·07
Cr_2O_3	151·97	SnO	134·68
CuO	79·53	SrO	103·61
F	19·00	Ta_2O_5	441·87
FeO	71·84	ThO_2	264·03
Fe_2O_3	159·68	TiO_2	79·89
H_2O	18·01	UO_2	270·02
HfO	210·48	U_3O_8	842·04
K_2O	94·20	V_2O_6	181·85
La_2O_3	325·80	Y_2O_3	225·79
Li_2O	29·87	ZnO	81·36
MgO	40·31	ZrO_2	123·21
MnO	·70·93		

Glossary

ACF diagram   Triangular composition–assemblage diagram especially useful for basic igneous rocks. The three components plotted are A $-$ $[Al_2O_3]$, C $-$ $[CaO]$, F $-$ $[FeO+MgO]$.

AFM diagram   Triangular composition–assemblage diagram especially useful for pelitic rocks. The three components plotted are A $-$ $[Al_2O_3]$, F $-$ $[FeO]$, M $-$ $[MgO]$.

age spectrum   A set of ages obtained from different fractions of argon driven off at progressively increasing temperatures in the potassium–argon spectrum age dating technique.

amphibolite   A metamorphic rock composed mainly of amphibole and plagioclase feldspar, usually with epidote and quartz.

argillite   A regional metamorphic rock of pelitic composition and low metamorphic grade, lacking **cleavage** or **schistosity**.

assemblage list   See **mineral assemblage**.

asthenosphere   The plastic layer in the Earth's mantle between 100 km and 700 km, in which the movements of the **lithospheric** plates are accommodated.

augen gneiss   A variety of **gneiss** whose gneissose banding wraps round porphyroblasts, giving the appearance of eyes.

autochthon   Untransported tectonic basement in an orogenic belt where there are transported **nappes**.

banded   A descriptive, non-genetic term for layers of alternating composition in metamorphic rocks. Used to draw a distinction from igneous layering and sedimentary bedding.

Barrow type metamorphic sequence   A regional metamorphic sequence developing first kyanite, then sillimanite with increasing **metamorphic grade** in pelitic rocks.

basic granulite   A name often applied to **pyroxene gneiss** of basic igneous composition.

blocking temperature   The temperature below which radiogenic argon is retained in minerals.

blueschist   A metamorphic rock of basic igneous composition containing glaucophane.

Buchan type metamorphic sequence   A regional metamorphic sequence developing first andalusite, then sillimanite with increasing **metamorphic grade** in pelitic rocks.

calc-schist   A **schist** containing carbonate minerals; a metamorphosed marl.

charnockite   The potassium-rich member of the **charnockitic suite**.

charnockitic gneiss   Metamorphic rock of acid igneous composition with gneissose banding, anhydrous mineral assemblage, granoblastic texture, and dark-coloured quartz due to the presence of rutile needles.

charnockitic suite Metamorphic rocks with granitic compositions (in the wide sense) but with only anhydrous minerals in their assemblages. Dark coloured in hand-specimen because of the presence of exsolved rutile needles in quartz.

chiastolite A variety of andalusite found in **contact aureoles**, containing opaque inclusions which form a cross in cross section.

cleavage A tendency to split along parallel planes, in both rocks and minerals. In rocks it should strictly be called 'rock cleavage'.

component Limiting chemical formula which may be used to express the composition of **phases** in a chemical **system**, e.g. the composition of a clinopyroxene phase in a pyroxene hornfels may be expressed as 68% of the component $CaMgSi_2O_6$ (diopside) and 32% of the component $CaFeSi_2O_6$ (hedenbergite). Note that 'component' is a general term which may also be used in chemical systems in which **diadochy** does not occur.

contact aureole The zone of **hornfelses** and other contact metamorphic rocks surrounding an igneous intrusion.

contact metamorphic rocks Metamorphic rocks formed against the contacts of igneous intrusions, assumed to have formed by baking due to the heat of the intrusion.

contact metasomatism **Metasomatism** occurring at the contacts of igneous intrusions.

continental accretion The theory that the continental areas of the Earth's crust have grown by the accretion of new continental crust at orogenic belts along their margins.

coronas Reaction rims between minerals formed by diffusion of ions across grain boundaries in the solid state.

country rocks Those rocks surrounding an igneous intrusion; sometimes also applied to those surrounding a mineral vein, fault or thrust.

crenulation cleavage A set of closely spaced parallel fracture surfaces which run parallel to the axial planes of microfolds of the pre-existing **schistosity** in a schistose metamorphic rock.

daughter isotopes Isotopes formed by radioactive decay, analysed in age dating.

dedolomitisation Breakdown of the magnesium-bearing component of dolomitic rocks during metamorphism.

degrees of freedom The number of ways in which a chemical **system** may be changed without altering the number or type of **phases** in the system.

deuteric alteration Secondary alteration of igneous rocks after solidification, usually with the formation of hydrous minerals. Generally regarded as an igneous process.

diadochy The substitution of one ion for another in the atomic structure of a mineral, without a change in structure, e.g. $Mg^{+2} \rightleftharpoons Fe^{+2}$ in olivine.

diagenesis Recrystallisation of rocks under conditions similar to those at or near the Earth's surface. Generally regarded as a sedimentary process.

directional fabric A metamorphic **texture** in which individual fabric elements, usually crystals of the metamorphic mineral assemblage, show a preferred orientation due to combined deformation and recrystallisation, especially in regional metamorphic rocks.

discontinuous reaction A metamorphic reaction in which the participating minerals have fixed compositions (i.e. **diadochy** does not occur).

divariant system A chemical **system** with two **degrees of freedom**. Represented by an area of a **phase diagram**.

dynamic metamorphic rocks Rocks found in or near faults, thrusts or meteorite impact sites, which have formed by intense deformation.

eclogite A rock of basic igneous composition, with the mineral assemblage garnet + omphacite.

enderbite The sodium-rich member of the **charnockitic suite**.

exsolution The separation of a single mineral **phase**, stable at high temperature, into two or more mineral phases as the rock cools.

fault breccia Fractured rock with angular fragments found in fault planes.

fault gouge Rock flour and fault breccia in a fault plane, partly altered to clayey material by weathering.

fibrolite Fine acicular or prismatic variety of sillimanite.

foliation Any metamorphic **texture** with a planar orientation, e.g. **slaty cleavage, gneissose banding**.

geothermal gradient The rate of increase of temperature with depth and therefore pressure inside the.Earth.

Gibbs free energy A thermodynamic parameter measuring the chemical potential energy of minerals participating in metamorphic reactions.

gneiss A banded metamorphic rock, with alternating bands rich and poor in ferromagnesian minerals.

gneissose banding (gneissosity) Metamorphic banding with an alternation of bands rich in quartz and feldspar with bands rich in ferromagnesian minerals.

Goldschmidt's mineralogical phase rule The number of minerals in a **mineral assemblage** is equal to, or less than, the number of oxide **components** in the rock analysis.

gram molecular weight The weight of one formula unit of a substance (e.g. a mineral or an oxide in a rock analysis), calculated from the atomic weight of each element multiplied by the number of atoms in the formula, and expressed in grammes. Often abbreviated to 'mole'.

granitisation A process whereby rocks are progressively transformed in the solid state, below their melting points, into granite. A special case of **regional metasomatism**.

granoblastic texture A texture formed of equant mineral grains of uniform grain size, with 120° triple junctions.

granofels A metamorphic rock of pelitic composition and high metamorphic grade, lacking a **directional fabric** in phyllosilicates.

granulite A quartz- and feldspar-rich metamorphic rock type from Saxony, Germany, with distinctive textures and mineral assemblages. The name is frequently extended to other metamorphic rocks (e.g. 'basic granulites').

greenschist A metamorphic rock of basic igneous composition with a distinct **cleavage**, containing abundant chlorite and of low metamorphic grade.

greenstone A metamorphic rock of basic igneous composition, lacking a **cleavage**. It typically contains actinolite, epidote and albite, and is therefore a variety of **amphibolite**.

grit Obsolete metamorphic rock name for semi-pelitic **schist**, metamorphosed greywacke or metamorphosed conglomerate.

half life   The time taken for half the unstable isotopes in a sample to decay. Radioactive isotopes used for age dating in metamorphic rocks have half lives of many millions of years.

halleflinta   A fine-grained, low grade metamorphic rock of acid igneous composition. Inclined to splinter when hammered.

hornfels   A massive, often flinty, contact metamorphic rock found nearest to the contact in **contact** aureoles.

illite crystallinity   A numerical indication of the ratio of illite to muscovite in sericites from low grade pelitic rocks. Used to measure **metamorphic grade** quantitatively in such rocks.

impactite glass   Glass formed by shock metamorphism and shock melting.

index minerals   Minerals marking definite stages (**metamorphic zones**) of increasing **metamorphic grade** in a **progressive metamorphic sequence**.

injection gneiss   Gneiss which has formed by the injection of granite magma parallel to the bedding of sedimentary rocks or the schistosity of metamorphic rocks. A type of migmatite.

invariant   A chemical system with no **degrees of freedom**. Represented by a point on a **phase diagram**.

isochron diagram   A plot of the ratios of parent and daughter isotopes to a stable isotope (e.g. $^{87}Rb/^{86}Sr$ and $^{87}Sr/^{86}Sr$). The slope of the resulting straight line graph gives the age of a rock (Fig. 14.5).

isograd   A surface joining points of equal **metamorphic grade** in a body of metamorphic rocks. The metamorphic grade is most often defined by the appearance of an **index mineral**. An isograd plots as a line on a geological map.

kimberlite   Fragmental volcanic rock containing fragments of rocks from deep in the Earth, sometimes including diamonds.

lineation   A metamorphic **texture** in which minerals or structures in the rock define a set of parallel lines (Fig. 2.9).

lithosphere   The outermost 100 km (approx.) of the solid Earth, including crust and mantle, which is rigid mechanically and lies above the more plastic **asthenosphere**.

marble   Metamorphic rock composed of carbonate minerals, especially calcite.

metamorphic differentiation   The separation of bands of rock of different composition by local **metasomatism** during high grade metamorphism.

metamorphic fabric   An alternative name for metamorphic **texture**. The term tends to be used for **directional fabrics**.

metamorphic facies   Subdivision of the conditions of metamorphism on the basis of diagnostic **mineral assemblages** or metamorphic reactions. Also used as the term for an individual subdivision recognised in this way.

metamorphic grade   A measure of the intensity of metamorphism, often relative. A rock nearer the intrusion in a contact metamorphic sequence is said to have a higher metamorphic grade than one further away.

metamorphic zone   A subdivision of a progressive metamorphic sequence, in contact or regional metamorphic rocks, recognised either by the incoming of characteristic **index minerals**, or by textural variations in the rocks.

metasomatism   Metamorphism which includes the addition to or removal from the rock of chemical **components**, other than volatile components such as H_2O or CO_2.

metastable A mineral is metastable when it is apparently stable, but is actually under conditions outside its stability range.

meteoric water line The straight line on a $\delta D/\delta^{18}O$ diagram obtained by plotting the isotopic compositions of natural water from rain or snow, which vary with climate (Fig. 14.8).

microstructure A synonym for **texture**, used when there may be confusion with metallurgical terms.

migmatite A mixed rock, often a **gneiss**, with granite or foliated granite mixed with high grade metamorphic rock.

mineral assemblage A list of minerals which coexisted in equilibrium during the metamorphism of a metamorphic rock.

mineral isochron A straight line on an isochron diagram obtained by plotting the different isotopic compositions of minerals in the same rock.

mole Abbreviation of **gram molecular weight**.

mortar texture Small crystals along the grain boundaries of earlier large ones in dynamic metamorphic rocks (Fig. 6.3).

mylonite A fine-grained, flinty dynamic metamorphic rock found in fault and thrust zones. Usually banded parallel to the fracture plane. Contains relict **porphyroclasts** of country rock.

nappe A sheet of rocks transported tectonically over a considerable distance above a flat or gently dipping thrust plane in an orogenic belt (especially the Alps).

non-penetrative fabric A metamorphic fabric composed of separate microstructures (such as close-spaced fractures).

nucleation The development of small 'seeds' from which crystals of new minerals can grow.

obduction Uplift and tectonic transport of ocean floor onto continental crust.

omphacite Sodium-rich variety of the clinopyroxene augite, found in blueschist and eclogite.

ophiolite suite Pillow lavas, gabbros and serpentinites from the Alpine orogenic belt. Often used as a synonym for 'fossil ocean floor'.

paragenesis An equivalent term to mineral **assemblage** in metamorphic rocks.

pelitic rock A metamorphosed sedimentary rock which was originally of shale composition.

penetrative fabric A **directional fabric** which influences every mineral grain in the rock (e.g. **slaty cleavage**).

peristerite Plagioclase feldspar in the composition range $An_8 - An_{17}$, which is a sub-microscopic intergrowth of albite and oligioclase.

petrogenesis The branch of petrology which discusses the mode of formation of rocks. A mode of formation proposed by this branch of petrology.

petrogenetic grid A diagram linking the stability ranges of mineral assemblages with several members to the conditions of metamorphism, not necessarily temperature and pressure alone, but activities of chemical components also.

phase A physically distinct substance in an experimental system (Ch. 4, the Phase Rule). Used loosely to describe a mineral species in a metamorphic **mineral assemblage**.

phase diagram A diagram representing the status of a chemical system in terms of the Phase Rule. If there are two **degrees of freedom** or less, a phase diagram can be drawn on two-dimensional paper. If temperature and pressure only are considered a phase diagram can show the ranges of stability of minerals during metamorphism.

Phase Rule A law of physical chemistry relating the number of **phases** in a chemical system to the number of chemical **components** and the **degrees of freedom**.

phengite A magnesium-rich variety of muscovite.

phyllite A metamorphic rock of pelitic composition, with a **cleavage**, coarser grained than **slate** and finer grained than **schist**.

phyllonite A slate-like rock, with cleavage parallel to the fault plane, formed by dynamic metamorphism.

plate tectonics A theory describing the movements of the Earth's outer 100 km, or **lithosphere**. This is taken to consist of a finite number of plates, which interact in different ways along narrow surfaces at their edges.

pleonaste Green variety of spinel, intermediate in composition between spinel ($MgAl_2O_4$) and hercynite ($FeAl_2O_4$).

poikiloblastic texture A texture in which large crystals of metamorphic minerals enclose smaller ones.

porphyroblast A large crystal of a metamorphic mineral in a finer grained groundmass.

porphyroblastic texture A texture with large crystals of metamorphic minerals in a finer grained groundmass.

porphyroclast Larger mineral fragment in a finer grained groundmass of a dynamic metamorphic rock, such as **mylonite**.

pressure solution Selective solution at certain points on the surface of mineral grains during deformation under low grade metamorphic conditions, or in sediments.

principle of uniformitarianism The belief that processes which may be observed changing the features of the Earth today formed rocks and geological structures in the past. Summed up in the phrase 'The present is the key to the past'.

progressive decarbonation Progressive loss of CO_2 from carbonate rocks with increasing metamorphic grade.

progressive dehydration Progressive loss of H_2O from hydrous rocks with increasing metamorphic grade.

progressive metamorphic sequence A series of metamorphic rocks in which the metamorphic grade increases steadily without breaks.

psammitic rock Metamorphosed sedimentary rock of sand or arkose composition.

pseudomorph A crystal which has been entirely replaced by a secondary mineral, but retains its original outline.

pseudotachylyte Glassy rock produced by frictional melting in a fault or thrust zone.

pyroxene gneiss Regional metamorphic rock of basic igneous composition displaying **gneissose banding**, with an anhydrous mineral assemblage and **granoblastic** texture.

pyroxene hornfels A type of **hornfels** commonly found in **contact aureoles**. Pelitic pyroxene hornfels contains dominant orthopyroxene (Fig. 5.8); basic igneous pyroxene hornfels, dominant clinopyroxene (Fig. 11.5).

quartzite A metamorphic rock composed mainly of quartz. A metamorphosed sandstone.

realistic reaction A chemical reaction which is convincingly demonstrated to have occurred in rocks during metamorphism.

regional metamorphic rocks Metamorphic rocks forming large parts of the Earth's crust and mantle, but with no obvious genetic association with igneous intrusions or major faults or thrusts.

regional metasomatism Regional metamorphism including the introduction or removal on a regional scale of material other than volatile components such as H_2O and CO_2.

regolith The fragmental surface layer of the Moon. Loosely speaking, lunar soil.

relict Relict minerals, textures or structures are survivors from the original igneous or sedimentary state of the rock, or from an earlier episode of metamorphism.

retrograde metamorphism Metamorphism in which an earlier high grade **mineral assemblage** is replaced or partially replaced by a later low grade assemblage.

saccharoidal marble Coarse-grained marble with **granoblastic** texture, which looks like sugar crystals.

schist Metamorphic rock of pelitic composition, with a well developed **schistosity**.

schistosity Planar **directional fabric** of medium- to coarse-grained phyllosilicates, commonly seen in schist.

semi-pelitic rock A metamorphosed sedimentary rock which was originally intermediate in composition between shales and sandstones.

sericite An informal name for white micas of the illite-muscovite series.

serpentinite A metamorphosed ultrabasic igneous rock, composed of minerals of the serpentine group.

shock metamorphism Metamorphism of rocks by intense shock waves produced by meteorite impacts and artificial nuclear explosions.

sieve texture A variety of **poikiloblastic texture** in which the poikiloblastic crystals have very many fine-grained inclusions.

skarn A rock containing one or two minerals only, formed by contact **metasomatism**. Often restricted to calcium-rich rocks at contacts with limestone and marble.

slate A pelitic metamorphic rock displaying **slaty cleavage**.

slaty cleavage (occasionally spelt 'slatey cleavage'). A penetrative **directional fabric** of the phyllosilicate minerals in slates, which enables the rock to be split into very thin, almost perfectly flat sheets.

sliding reaction A metamorphic reaction involving two or more minerals in which **diadochy** of reacting components occurs. The reaction takes place over a range of temperature and pressure, not at a **univariant** curve.

smectite A general name for clay minerals of the montmorillonite group. Used in Chapter 11 for clay minerals formed by low grade hydration of basaltic glass.

SMOW ('Standard Mean Ocean Water') The standard isotopic composition of water used in hydrogen and oxygen isotope studies.

soapstone A massive metamorphosed ultrabasic rock, composed largely of talc.

solid solution A synonym for **diadochy**. One end-member component of a mineral is described as 'dissolving' in another e.g. Fe_2SiO_4 dissolving in Mg_2SiO_4 in olivines.

spotted slate A contact metamorphic rock, usually of pelitic composition, with dark spots appearing on **slaty cleavage** surfaces (which formed during earlier regional metamorphism).

strain–slip cleavage A synonym for **crenulation cleavage**.

structural petrology   The analysis of the geometrical and genetic relationships between major geological structures, minor geological structures and **metamorphic fabrics**.

suevite   A breccia of rock fragments in impactite glass.

system   An experimental or theoretical isolated sample of material of simple chemical composition, used for the study of the mutual stability relationships of individual **phases**. Usually described in terms of its limiting chemical formulae, e.g. the system $MgO–SiO_2–H_2O$.

talc schist   A metamorphosed ultrabasic rock with a **schistosity**, largely composed of talc.

texture   The relationship of size, shape and orientation between the mineral grains of a metamorphic rock. The term **microstructure** has recently come into use with the same meaning.

thermal metamorphism   Metamorphic recrystallisation caused by heating to a high temperature without contemporary deformation. Very commonly used as a synonym for **contact metamorphism**.

tie-line   A line on a composition–assemblage diagram joining the compositions of minerals coexisting in equilibrium.

triple point   A point on a **phase diagram** where three curves representing **univariant** states of a chemical **system** meet. Represents an **invariant** state of the system.

ultramylonite   Mylonite in which there are very few or no **porphyroclasts**.

univariant   A chemical system with one **degree of freedom**, represented by a curve on a **phase diagram**. Discontinuous metamorphic reactions are represented by univariant curves on temperature–pressure phase diagrams.

whole rock isochron   An isochron plotted by using several rocks of different isotopic composition.

xenolith   A block of country rock surrounded by igneous rock in an intrusion.

References

Agrell, S. O. 1965. Polythermal metamorphism of limestones at Kilchoan, Ardnamurchan. *Mineral. Mag.* **34** (Tilley Vol.), 1–15.

Anderson, D. E. and G. R. Buckley 1973. Zoning in garnets — diffusion models. *Contr. Mineral. Petrol.* **40**, 87–104.

Atherton, M. P. 1968. The variation in garnet, biotite and chlorite composition in medium grade pelitic rocks from the Dalradian, Scotland, with particular reference to the zonation in garnet. *Contr. Mineral. Petrol.* **18**, 347–71.

Ayrton, S. 1969. On the origin of gneissose banding. *Eclogae Geol. Helveticae* **62**, 567–76.

Badley, M. E. 1976. Stratigraphy, structure and metamorphism of Dalradian rocks of the Maumturk Mountains, Connemara, Ireland. *J. Geol. Soc. Lond.* **132**, 509–20.

Bailey, E. B. 1935. *Tectonic essays: mainly Alpine.* Oxford: Oxford University Press (facsimile reprint 1968).

Barrow, G. 1912. On the geology of lower Dee-side and the southern Highland border. *Proc. Geol. Assoc.* **23**, 274–90.

Battey, M. H. 1972. *Mineralogy for students.* Edinburgh: Oliver & Boyd.

Binns, R. A. 1965. The mineralogy of metamorphosed basic rocks from the Willyama complex, Broken Hill district, New South Wales. *Mineral. Mag.* **35**, 306–26.

Boullier, A. M. and A. Nicholas 1975. Classification of textures and fabrics of peridotite xenoliths from South African kimberlites. *Phys. and Chem. of the Earth* **9**, 467–75.

Bowes, D. R. 1976. Archaean crustal activity in north-western Britain. In *The early history of the Earth*, B. F. Windley (ed.), 469–79. London: John Wiley.

Bowes, D. R., A. E. Wright and R. G. Park 1964. Layered intrusive rocks in the Lewisian of the North-West Highlands of Scotland. *Q. J. Geol. Soc. Lond.* **120**, 153–84.

Carmichael, I. S. E., F. J. Turner and J. Verhoogen 1974. *Igneous petrology.* New York: McGraw-Hill.

Chinner, G. A. 1966. The distribution of temperature and pressure during Dalradian metamorphism. *Q. J. Geol. Soc. Lond.* **122**, 159–86.

Chinner, G. A. and J. E. Dixon 1973. Some high pressure parageneses of the Allalin Gabbro, Valais, Switzerland. *J. Petrol.* **14**, 185–202.

Cooper, A. F. 1972. Progressive metamorphism of metabasic rocks from the Haast Schist Group of Southern New Zealand. *J. Petrol.* **13**, 457–92.

Dallmeyer, R. D. 1975. $^{40}Ar/^{39}Ar$ ages of biotite and hornblende from a progressively remetamorphosed basement terrane: their bearing on the interpretation of release spectra. *Geochim. et Cosmochim. Acta* **39**, 1655–69.

Dalrymple, G. B. and M. A. Lanphere 1969. *Potassium–argon dating.* San Francisco: Freeman.

Davies, G. R. 1969. Aspects of the metamorphosed sulphide ores at Kilembe, Uganda. In *Sedimentary ores ancient and modern (revised)*, C. H. James (ed.), 273–95. Proc. 15th Inter-University Geological Congress, Leicester 1967.

Deer, W. A., R. A. Howie and J. Zussman 1966. *Introduction to the rock-forming minerals*. London: Longman.

Dewey, J. F. 1969. The evolution of the Appalachian/Caledonian orogen. *Nature* (London) **222**, 124–9.

Dewey, J. F. and J. M. Bird 1970. Mountain belts and the new global tectonics. *J. Geoph. Res.* **75**, 2625–47.

Dewey, J. F. and R. J. Pankhurst 1970. Evolution of the Scottish Highlands and their radiometric age pattern. *Trans. Roy. Soc. Edinb.* **68**, 361–89.

Dewey, J. F., W. S. McKerrow and S. Moorbath 1970. The relationship between isotopic ages, uplift and sedimentation during Ordovician times in western Ireland. *Scottish J. Geol.* **6**, 133–45.

Dorn, P. 1960. *Geologie von Mitteleuropa*. Stuttgart: Schweizerbart'sche.

Dunoyer de Segonzac, G. 1970. The transformation of clay minerals during diagenesis and low-grade metamorphism: a review. *Sedimentology* **15**, 281–346.

Eastwood, T., S. E. Hollingworth, W. C. C. Rose and F. M. Trotter 1968. *Geology of the country around Cockermouth and Caldbeck. (Explanation of One-inch Geological Sheet 23, New Series)*. London: Institute of Geological Sciences.

Eskola, P. 1915. *On the relations between the chemical and mineralogical composition in the metamorphic rocks of the Orijärvi region*. Bull. Comm. Géol. Finlande, no. 44.

Eskola, P. 1920. The mineral facies of rocks. *Norsk Geol. Tidsskr.* **6**, 143–94.

Evans, A. M., T. D. Ford and J. R. L. Allen 1968. Precambrian rocks. In *Geology of the East Midlands*, P. C. Sylvester-Bradley and T. D. Ford (eds), 1–19. Leicester: Leicester University Press.

Faure, G. and J. L. Powell 1972. *Strontium isotope geochemistry*. Heidelberg: Springer-Verlag.

Ferguson, J., H. Martin, L. O. Nicolaysen and R. V. Danchin 1975. Gross Brukkaros: a kimberlite–carbonatite volcano. *Phys. and Chem. of the Earth* **9**, 219–34.

Frey, M. 1974. Alpine metamorphism of pelitic and marly rocks of the Central Alps. *Schweiz. Mineral. und Petrogr. Mitt.* **54**, 489–506.

Frey, M. and J. C. Hunziker 1973. Progressive niedriggradige Metamorphose glaukonitführender Horizonte in den helvetischen Alpen der Ostschweiz. *Contr. Mineral. Petrol.* **39**, 185–218.

Frey, M., J. C. Hunziker, W. Frank, J. Bocquet, G. V. Dal Piaz, E. Jäger and E. Niggli 1974. Alpine metamorphism of the Alps — A review. *Schweiz. Mineral. und Petrol. Mitt.* **54**, 247–90.

Fyfe, W. S., F. J. Turner and J. Verhoogen 1958. *Metamorphic reactions and metamorphic facies*. Geol. Soc. America, Memoir 73.

Garlick, G. D. and S. Epstein 1967. Oxygen isotope ratios in coexisting minerals of regionally metamorphosed rocks. *Geochim. et Cosmochim. Acta* **31**, 181–214.

Gass, I. G. and J. D. Smewing 1973. Intrusion, extrusion and metamorphism at constructive margins: evidence from the Troodos Massif, Cyprus. *Nature* (London) **242** 26–9.

Gass, I. G., P. J. Smith and R. C. L. Wilson 1971. *Understanding the Earth: a reader in the earth sciences*. Horsham, Sussex: Artemis Press.

Goldschmidt, V. M. 1911. *Die Kontaktmetamorphose im Kristianiagebiet*. Vidensk. Skrifter I. Mat.-Naturv. K. (1911), No. 11.

Hall, R. 1976. Ophiolite emplacement and the evolution of the Taurus suture zone, southeastern Turkey. *Bull. Geol. Soc. America* **87**, 1078–88.

Hanson, G. N., K. R. Simmons and A. E. Bence 1975. $^{40}Ar/^{39}Ar$ spectrum ages for biotite, hornblende and muscovite in a contact metamorphic zone. *Geochim. et Cosmochim. Acta* **39**, 1269–77.

Harker, A. 1932. *Metamorphism*. London: Methuen.

Hart, S. R. 1964. The petrology and isotopic-mineral age relations of a contact zone in the Front Range, Colorado. *J. Geol.* **72**, 493–525.

Harte, B. 1975. Determination of a pelite petrogenetic grid for the eastern Scottish Dalradian. *Carnegie Inst. Washington, Yearbook* **74**, 438–46.

Harte, B. and J. J. Gurney 1975. Evolution of clinopyroxene and garnet in an eclogite nodule from the Roberts Victor kimberlite pipe, South Africa. *Phys. and Chem. of the Earth* **9**, 367–87.

Harte, B. and K. J. Henley 1966. Zoned almanditic garnets from regionally metamorphosed rocks. *Nature* (London) **210**, 689–92.

Hartshorne, N. H. and A. Stuart 1950. *Crystals and the polarising microscope.* London: Edward Arnold.

Hatch, F. H., A. K. Wells and M. K. Wells 1972. *Petrology of the igneous rocks.* London: George Allen & Unwin.

Hawthorne, J. B. 1975. Model of a kimberlite pipe. *Phys. and Chem. of the Earth* **9**, 1–15.

Hedberg, H. D. (ed.) 1976. *International stratigraphic guide*. New York: John Wiley.

Heinrich, E. W. 1965. *Microscopic identification of minerals*. New York: McGraw-Hill.

Henley, K. J. 1970. The structural and metamorphic history of the Sulitjelma region, Norway, with special reference to the nappe hypothesis. *Norsk Geol. Tidssk.* **50**, 97–136.

Hensen, B. J. and D. H. Green 1970. Experimental data on coexisting cordierite and garnet under high grade metamorphic conditions. *Phys. Earth Planet. Interiors* **3**, 431–40.

Hobbs, B. E., W. D. Means and P. F. Williams 1976. *An outline of structural geology*. New York: John Wiley.

Hodges, C. A., W. R. Muehlenberger and G. E. Ulrich 1973. Geologic setting of Apollo 16. *Proc. Fourth Lunar Sci. Conf.* (Supplement 4 of *Geochim. et Cosmochim. Acta*), Vol. 1, 1–25.

Hollister, L. S. 1966. Garnet zoning: an interpretation based on the Rayleigh fractionation model. *Science* (N.Y.) **154**, 1647–51.

Holmes, A. 1965. *Principles of physical geology*. London: Nelson.

Howie, R. A. 1955. The geochemistry of the charnockite series of Madras, India. *Trans. Roy. Soc. Edinburgh* **62**, 725–68.

Kerr, P. F. 1959. *Optical mineralogy*, 3rd ed., New York: McGraw-Hill.

Lappin, M. A. and J. B. Dawson 1975. Two Roberts Victor eclogites and their re-equilibration. *Phys. and Chem. of the Earth* **9**, 351–66.

Leake, B. E. 1958. Composition of pelites from Connemara, Co. Galway, Ireland. *Geol. Mag.* **101**, 63–75.

Leake, B. E. 1970a. The origin of the Connemara migmatites of the Cashel district, Connemara, Ireland. *Q. J. Geol. Soc. London* **125**, 219–76.

Leake, B. E. 1970b. The fragmentation of the Connemara basic and ultrabasic intrusions. In *Mechanisms of igneous intrusions*, G. Newall and N. Rast (eds), 103–22. Liverpool: Liverpool Geological Society.

Long, J. V. P. 1967. Electron-probe microanalysis. In *Physical methods in determinative mineralogy*, J. Zussman (ed.), 215–60. London: Academic Press.

Macgregor, M. 1972. *Excursion guide to the geology of Arran*, 2nd ed. Glasgow: Geological Society of Glasgow.

Mason, R. 1967. Electron-probe microanalysis of coronas in a troctolite from Sulitjelma, Norway. *Mineral. Mag.* **36**, 504–14.

Mason, R. 1971. The chemistry and structure of the Sulitjelma gabbro. *Norges geol. Unders.* no. 269, 108–41.

Mason, R. 1975. The tectonic status of the Bitlis Massif. *Proc. 50th Anniversary Congress of Earth Science, Ankara 1973*, 31–41.

Mehnert, K. R. 1968. *Migmatites and the origin of granitic rocks.* Amsterdam: Elsevier.

Mercy, E. L. P. 1965. Caledonian Igneous Activity. In *Geology of Scotland*, G. Y. Craig (ed.), 229–67. Edinburgh: Oliver & Boyd.

Miyashiro, A. 1961. Evolution of metamorphic belts. *J. Petrol.* **2**, 277–311.

Miyashiro, A. 1973. *Metamorphism and metamorphic belts.* London: George Allen & Unwin.

Miyashiro, A., F. Shido and M. Ewing 1971. Metamorphism in the Mid-Atlantic Ridge near 24° and 30° N. *Phil. Trans. Roy. Soc. London*, Ser. A. **268**, 589–603.

Moores, E. M. 1969. *Petrology and structure of the Vourinos ophiolite complex of northern Greece.* Geol. Soc. America, Special Paper 118.

Moores, E. M. and F. J. Vine 1971. The Troodos Massif, Cyprus and other ophiolites as oceanic crust: evolution and implications. *Phil. Trans. Roy. Soc. London*, Ser. A. **268**, 443–66.

Mutch, T. A. 1972. *Geology of the Moon: a stratigraphic view.* Princeton N. J.: Princeton University Press.

Naggar, M. H. and M. P. Atherton 1970. The composition and metamorphic history of some aluminium silicate-bearing rocks from the aureoles of the Donegal granites. *J. Petrol.* **11**, 549–89.

O'Hara, M. J. 1961. Zoned ultrabasic and basic gneiss masses in the Early Lewisian Complex at Scourie, Sutherland. *J. Petrol.* **2**, 248–76.

Oxburgh, E. R. 1974. The plain man's guide to plate tectonics. *Proc. Geol. Assoc.* **85**, 299–357.

Oxburgh, E. R. and D. L. Turcotte 1971. Origin of paired metamorphic belts and crustal dilation in island-arc regions. *J. Geophys. Res.* **76**, 1315–27.

Oxburgh, E. R. and D. L. Turcotte 1974. Thermal gradients and regional metamorphism in overthrust terrains with special reference to the eastern Alps. *Schweiz. Mineral. und Petrogr. Mitt.* **54**, 641–62.

Porteous, W. G. 1973. Metamorphic index minerals in the Eastern Dalradian. *Scottish J. Geol.* **9**, 29–43.

Ramsay, J. G. 1967. *Folding and fracturing of rocks.* New York: McGraw-Hill.

Rastall, R. H. 1910. The Skiddaw granite and its metamorphism. *Q. J. Geol. Soc. London* **66**, 116–41.

Read, H. H. 1952. Metamorphism and migmatisation in the Ythan valley, Aberdeenshire. *Trans. Edinburgh Geol. Soc.* **15**, 265–79.

Read, H. H. 1957. *The granite controversy.* London: Thomas Murby.

Read, H. H. and J. Watson 1975a. *Introduction to geology.* Volume 2: *Earth history.* Part I: *Early stages of Earth history.* London: Macmillan.

Read, H. H. and J. Watson 1975b. *Introduction to geology.* Volume 2: *Earth history.* Part II: *Later stages of Earth history.* London: Macmillan.

Reid, A. M., C. H. Donaldson, J. B. Dawson, R. W. Brown and W. I. Ridley 1975. The Igwisi Hills extrusive 'kimberlites'. *Phys. and Chem. of the Earth* **9**, 199–218.

Richardson, S. W., M. C. Gilbert and P. M. Bell 1969. Experimental determination of kyanite-andalusite and andalusite-sillimanite equilibria; the aluminium silicate triple point. *Am. J. Sci.* **267**, 259–72.

Robinson, D. 1972. Metamorphic rocks. In *Geology of Durham County*, G. Hickling (ed.), 119–23. Durham: Natural History Society of Northumberland, Durham and Newcastle upon Tyne.

Schmid, S. M. 1975. The Glarus overthrust: field evidence and mechanical model. *Eclogae geol. Helveticae* **68**, 247–80.

Skevington, D. 1971. Palaeontological evidence bearing on the age of Dalradian deformation and metamorphism in Ireland and Scotland. *Scottish J. Geol.* **7**, 285–8.

Smith, A. G. 1976. Plate tectonics and orogeny: a review. *Tectonophysics* **33**, 215–85.

Sobolev, V. S. (ed.) 1972. *The facies of metamorphism* (translated by D. A. Brown), Canberra: Australian National University. (Original Russian edition 1970. Moscow: Nedra.)

Spooner, E. T. C., R. D. Beckinsale, W. S. Fyfe and J. D. Smewing 1974. ^{18}O enriched ophiolitic metabasic rocks from E. Liguria (Italy), Pindos (Greece), and Troodos (Cyprus). *Contr. Mineral. Petrol.* **47**, 41–62.

Spry, A. 1969. *Metamorphic textures*. Oxford: Pergamon Press.

Stewart, F. H. 1965. Tertiary igneous activity. In *Geology of Scotland*, G. Y. Craig (ed.), 417–65. Edinburgh: Oliver & Boyd.

Stöffler, D. 1966. Zones of impact metamorphism in the crystalline rocks of the Nördlinger Ries Crater. *Contr. Mineral. Petrol.* **12**, 15–24.

Sutton, J. and J. Watson 1951. The pre-Torridonian metamorphic history of the Loch Torridon and Scourie areas in the North-west Highlands, etc. *Q. J. Geol. Soc. London* **106**, 241–307.

Taylor, H. P. 1974. The application of oxygen and hydrogen isotope studies to problems of hydrothermal alteration and ore deposits. *Econ. Geol.* **69**, 843–83.

Taylor, H. P. and R. W. Forester 1971. Low O^{18} igneous rocks from the intrusive complexes of Skye, Mull and Ardnamurchan, western Scotland. *J. Petrol.* **12**, 465–97.

Taylor, S. R. 1975. *Lunar Science: a post-Apollo view*. New York: Pergamon Press.

Thompson, J. B. 1957. The graphical analysis of mineral assemblages in pelitic schists. *Am. Mineral.* **42**, 842–58.

Tilley, C. E. 1924a. Contact metamorphism in the Comrie area of the Perthshire Highlands. *Q. J. Geol. Soc. London* **80**, 22–71.

Tilley, C. E. 1924b. The facies classification of metamorphic rocks. *Geol. Mag.* **61**, 167–71.

Tilley, C. E. 1925. A preliminary survey of metamorphic zones in the southern Highlands of Scotland. *Q. J. Geol. Soc. London* **81**, 100–12.

Tilley, C. E. 1926. On garnet in pelitic contact zones. *Mineral. Mag.* **21**, 47–50.

Tilley, C. E. 1936. Enderbite, a new member of the Charnockite Series. *Geol. Mag.* **73**, 312–16.

Tilley, C. E. 1951. The zoned contact skarns of the Broadford area, Skye. *Mineral. Mag.* **29**, 621–66.

Turner, F. J. 1968. *Metamorphic petrology*. New York: McGraw-Hill.

Turner, F. J. and L. E. Weiss 1963. *Structural analysis of metamorphic tectonites*. New York: McGraw-Hill.

Umbgrove, J. H. F. 1950. *Symphony of the Earth*. The Hague: Martinus Nijhoff.

Vernon, R. H. 1976. *Metamorphic processes*. London: George Allen & Unwin.

Vogt, Th. 1927. *Sulitelmafeltets geologi og petrografi*. Norges geol. Unders., no. 121.

Wager, L. R. and G. M. Brown 1968. *Layered igneous rocks*. Edinburgh: Oliver & Boyd.

Warner, J. L. 1972. Metamorphism of Apollo 14 breccias. *Proc. Third Lunar Sci. Conf.* (Supplement 3 to *Geochim. et Cosmochim. Acta*), Vol. 1, 623–43.

Warner, J. L., C. H. Simonds and W. C. Phinney 1973. Apollo 16 rocks: classification and petrogenetic model. *Proc. Fourth Lunar Sci. Congr.* (Supplement 4 to *Geochim. et Cosmochim. Acta*), Vol. 1, 481–504.

Wilson, M. R. 1971. Dating rocks from the Earth and Moon. In *The Earth and its Satellite*, J. E. Guest (ed.), 149–56. London: Rupert Hart-Davies.

Winchester, J. A. 1974. The zonal pattern of regional metamorphism in the Scottish Caledonides. *J. Geol. Soc. London* **130**, 509–25.

Winkler, H. G. F. 1965. *Petrogenesis of metamorphic rocks*, 1st edn. New York: Springer-Verlag.

Winkler, H. G. F. 1976. *Petrogenesis of metamorphic rocks*, 4th edn. New York: Springer-Verlag.

Wood, B. J. 1973. $Fe^{2+} - Mg^{2+}$ partition between coexisting cordierite and garnet – a discussion of the experimental data. *Contr. Mineral. Petrol.* **40**, 253–8.

Wood, B. J. and D. G. Fraser, 1977. *Elementary thermodynamics for geologists*. London: Oxford University Press.

Wyllie, P. J. 1971. *The dynamic Earth*. New York: John Wiley.

Yardley, B. W. D. 1976. Deformation and metamorphism of Dalradian rocks and the evolution of the Connemara cordillera. *J. Geol. Soc. London* **132**, 521–42.

Yoder, H. S. and C. E. Tilley 1962. Origin of basaltic magmas: an experimental study of natural and synthetic rock systems. *J. Petrol.* **3**, 342–532.

Index

Entries in **bold type** are in the glossary. Page references in **bold type** refer to figures and tables.